A Centrifugal Particle Receiver for High-Temperature Solar Applications

Von der Fakultät für Maschinenwesen der Rheinisch-Westfälischen Technischen Hochschule Aachen zur Erlangung des akademischen Grades einer Doktorin der Ingenieurwissenschaften genehmigte Dissertation

vorgelegt von

Wei Wu

Berichter: Univ.-Prof. Dr.-Ing. R. Pitz-Paal
Univ.-Prof. Dr.-Ing. R. Kneer

Tag der mündlichen Prüfung: 29.10.2014

Bibliografische Information der Deutschen Nationalbibliothek

Die Deutsche Nationalbibliothek verzeichnet diese Publikation in der Deutschen Nationalbibliografie; detaillierte bibliografische Daten sind im Internet über http://dnb.d-nb.de abrufbar.

D82 (Diss. RWTH Aachen University, 2014)

ISBN 978-3-8325-3882-8

Logos Verlag Berlin GmbH
Comeniushof, Gubener Str. 47,
10243 Berlin
Tel.: +49 (0)30 42 85 10 90
Fax: +49 (0)30 42 85 10 92
INTERNET: http://www.logos-verlag.de

"I'd put my money on the sun and solar energy.
What a source of power! I hope we don't have to
wait until oil and coal run out before we tackle that."

[Thomas Edison, 1931]

Danksagung

Ich möchte mich an dieser Stelle zunächst bei Herrn Professor Pitz-Paal für die Betreuung dieser Promotion bedanken. Seine sehr hilfreichen Anregungen und Ratschläge haben mich stets unterstützt und voran gebracht. Herrn Professor Kneer danke ich für das zweite Gutachten sowie seinen wertvollen Anmerkungen zu meiner Dissertation.

Mein herzlichster Dank gilt meinen Kolleginnen und Kollegen der SF in Stuttgart, die mit konstruktiven Beiträgen und fruchtbaren Diskussionen meine Arbeit unheimlich bereichert haben. Besonders Reiner und Lars bin ich zu tiefstem Dank verpflichtet, da sie mir stets mit gezielter Anleitung, inspirierenden Ratschlägen und unermüdlicher Geduld zur Seite standen. Die erfolgreiche Durchführung meiner Experimente im Hochleistungsstrahler wäre mir nicht gelungen, wenn ich nicht die unentbehrliche Unterstützung des dortigen Teams gehabt hätte. Ich danke Christian Willsch, Gerd Dibowski und Christian Räder für ihren Einsatz und ihrer völlig unkomplizierten Art die Sachen auch einfach mal anzupacken. Außerdem bin ich sehr dankbar für die tatkräftige Unterstützung einer Reihe von mir betreuter Studenten. Ohne sie wäre ich lange nicht soweit gekommen.

Schließlich gilt mein allertiefster Dank meinen Freunden, meiner Familie und Daniel. Sie haben mich in diesen letzten Jahren auf jede erdenkliche Art unterstützt und standen mir immer zur Seite.

Vielen Dank.

Zusammenfassung

Die Zielsetzung der vorliegenden Dissertation ist die experimentelle und numerische Untersuchung eines sogenannten Zentrifugalpartikelreceivers (CentRec), der im Rahmen dieser Arbeit für die kommerzielle Produktion von Strom und industrieller Hochtemperaturprozesswärme mit konzentrierenden, solarthermischen Kraftwerken entwickelt worden ist. Dank der einfachen und robusten Bauweise des Receivers sind insgesamt geringe Systemkosten zu erwarten, die bei hohen Solaranteilen und Jahreserträgen zu geringen Wärme- und Stromgestehungskosten führen. Nahezu schwarze, die konzentrierte Solarstrahlung direkt absorbierende Keramikpartikel dienen sowohl als Wärmeträger- als auch als Speichermedium und können bis zu 900 bis 1000 °C aufgeheizt werden. Die einfache und direkte Speichermöglichkeit der Partikel ermöglicht einen konstanten 24h-Betrieb einhergehend mit garantierter Verfügbarkeit und Prozessstabilität.

Das Konzept von CentRec basiert im Grunde auf einen rotierenden Hohlraumzylinder, der in verschiedenen Winkeln zur Horizontalen geneigt werden kann. Partikelmassenstrom ($\dot{m}$) und Rotationsgeschwindigkeit (Ω) können dabei so variiert werden, dass sich zu jeder Zeit bzw. bei jedem Lastzustand ein optisch dichter, aber dünner Partikelfilm an der Zylinderinnenwand ausbildet, der von der eintretenden Solarstrahlung sukzessiv aufgeheizt wird. Außerdem ist es möglich über $\dot{m}$ und Ω die Partikelaufenthaltsdauer so zu regeln, dass die Partikelaustrittstemperatur unabhängig von der eingestrahlten Leistung konstant auf die angestrebte Prozesstemperatur gehalten werden kann.

Während ein erster Prototyp des Receivers im Labormaßstab (15 kW) für die Demonstration der generellen Machbarkeit und die Bestimmung des thermischen Wirkungsgrades konstruiert und experimentell getestet worden ist, dient das parallel entwickelte, numerische Modell der detailierten Untersuchung der einzelnen Wärmeverlustmechanismen.

Erste Versuche ohne Bestrahlung der Partikel sollten die Kontrollierbarkeit des Partikelfilms in Abhängigkeit von Massenstrom und Rotationsgeschwindigkeit zeigen, wobei mit Hilfe einer High-Speed Kamera die Partikelbewegung aufgenommen und ausgewertet wurde. Es hat sich herausgestellt, dass passend zu jedem eingestellten Massenstrom eine kritische Rotationsgeschwindigkeit existiert, ab der sich erst ein optisch dichter und trotzdem dünner, sich bewegender Partikelfilm ausbilden kann. Die Bestimmung der Partikelaufenthaltsdauer im Receiver für verschiedene $\dot{m}$ und Ω, hat außerdem bewiesen, dass diese sich tatsächlich über Massenstrom und Rotationsgeschwindigkeit kontrollieren lässt.

Um das Potential von CentRec für solare Hochtemperaturanwendungen zu demonstrieren, wurden Experimente im Hochleistungsstrahler des DLRs in Köln durchgeführt, bei denen die Partikel auf eine Endtemperatur von 900 °C aufgeheizt und der thermische Wirkungsgrad bestimmt werden sollte. Verschiedene Einflussparameter, wie Einstrahlleistung, Neigungswinkel, Massenstrom und Rotationsgeschwindigkeit wurden dabei detailliert untersucht. Die angestrebte Partikelauslasstemperatur von 900 °C ist bei einem Massenstrom von $8\,\mathrm{g\,s^{-1}}$ und einer Einstrahlflussdichte von $670\,\mathrm{kW\,m^{-2}}$ erreicht worden. Der thermische Wirkungsgrad hierbei beträgt etwa 75 %. Dieser Fall stellt jedoch nur einen Teillastzustand dar, da aufgrund von Alterung der Lampen sowie Positionierungsungenauigkeiten nur etwa Zweidrittel der Auslegungsflussdichte von $1\,\mathrm{MW\,m^{-2}}$ des Receivers experimentell erreicht worden ist.

Da die Experimente keinen detaillierten Einblick in die einzelnen Verlustmechanismen des Receivers geben, wurde ein auf der Finiten-Elemente-Methode basiertes numerisches Modell von CentRec entwickelt, das mit den Versuchsdaten validiert wurde. Es konnte für alle untersuchten Parameter eine gute Übereinstimmung in Hinblick auf Partikelauslasstemperatur und absorbierte Leistung erzielt werden. Wie zu erwarten, machen Strahlungsverluste mit 22 % der Gesamtleistung bei Auslasstemperaturen von 900 °C den größten Anteil aus. Konvektionsverluste hingegen sind stark von der Receiverneigung abhängig und sind bei einem Neigungswinkel von 45° weitaus größer als bei 90°. Ähnlich wie die Leitunsgverluste, können Konvektionsverluste maximal 10 % der Gesamtleistung ausmachen. Optische Verluste aufgrund von Reflektionen spielen eher eine untergeordnete Rolle. Desweiteren konnte mit dem validierten Modell eine Leistungslastkurve des Receivers erstellt werden, die für die spätere Auslegungen realer Receiver von Nutzen ist. Das vielversprechende Potential von CentRec konnte an Hand der Modellberechnungen bestätigt werden, da sich für eine Partikelauslasstemperatur von 900 °C und einer Auslegungsleistung des Receivers von 15 kW (entspricht $1\,\mathrm{MW\,m^{-2}}$) ein thermischer Wirkungsgrad von $> 85\,\%$ ergibt.

Abstract

The novel concept of a particle receiver for high-temperature solar applications was developed and evaluated in the present work. The so-called Centrifugal Particle Receiver (CentRec) uses small bauxite particles as absorber, heat transfer and storage media at the same time. Due to advantageous optical and thermal properties, the particles can be heated up to 1000 °C without sintering in the storage. High thermal efficiencies at high outlet temperatures are expected indicating a promising way for cost reduction in solar power tower applications.

CentRec basically consists of a cylindrical cavity, that is exposed to a defined rotation. Particles of about 1 mm diameter are forced against the wall, where they form a thin, but optically dense particle film. The film is gradually heated by direct radiation and depending on the incoming heat flux, the particle residence time can be adjusted by controlling rotation speed and mass flow rate. A constant particle outlet temperature for all load conditions can be therefore ensured.

A prototype in laboratory scale (15 kW_{th}) was designed, built and tested in order to demonstrate the feasibility and potential of the proposed concept. Extensive experiments regarding the particle controllability and high flux tests for various power levels and mass flow rates were conducted and evaluated. The expected simple control capability of the receiver could be verified and the target outlet temperature of 900 °C successfully demonstrated. A numerical model of the CentRec prototype, which has been validated by measured data, supplements the experimental findings. Thermal receiver efficiencies of $> 85\,\%$ for a design power of $1\,MW\,m^{-2}$ and an outlet temperature of 900 °C are predicted.

Contents

Nomenclature

Roman Symbols

A area, [m]

a absorptivity, [-]

c, c_p heat capacity, $[\mathrm{J\,kg^{-1}\,K^{-1}}]$

D diameter, [m]

$d\dot{Q}$ relative heat flow rate, [-]

F view factor, [-]

g gravitational constant, $[\mathrm{m\,s^{-2}}]$

H lead, [m]

h heat transfer coefficient, $[\mathrm{W\,m^{-1}\,K^{-2}}]$

L length, [m]

$\dot{m}$ mass flow rate, $[\mathrm{kg\,s^{-1}}]$

$[K], [L], [U]$ finite element matrices

$\boldsymbol{n}$ normal surface vector, [-]

N number, [-]

Nu	Nusselt number, [-]
P	electrical power, [W]
Pr	Prandtl number, [-]
$\dot{Q}$	heat flow rate, [W]
$\dot{q}$	heat flux, [$\mathrm{W\,m^{-2}}$]
q	quantity of interest
R	radius, [m]
r	distance between radiative surfaces
Ra	Rayleigh number, [-]
Ro	Rossby number, [-]
S	surface area, [$\mathrm{m^2}$]
$\dot{S}$	internal heat generation rate per unit volume, [$\mathrm{W\,m^{-3}}$]
T	temperature, [K]
t	time, [s]
Ta	Taylor number, [-]
u, v, w	velocity components, [$\mathrm{m\,s^{-1}}$]
x, y, z	directions in kartesian coordinate system, [-]

Greek Symbols

α	receiver inclination angle, [°]
β	thermal expansion coefficient, [$\mathrm{K^{-1}}$]
χ	helix angle, [-]

δ	uncertainty, thickness [m]
δ_p	particle layer thickness, [mm]
ε	emissivity, [-]
η	receiver efficiency, [-]
κ	thermal diffusivity, $[\mathrm{m^2\,s^{-1}}]$
δ_{ij}	Kronecker delta
λ	thermal conductivity, $[\mathrm{W\,m^{-1}\,K^{-1}}]$
ν	kinematic viscosity, $[\mathrm{m^2\,s^{-1}}]$
Ω	angular rotation speed, $[\mathrm{s^{-1}}]$
φ	circumferential direction
Ψ	arbitrary variable
ρ	density $[\mathrm{kg\,m^{-3}}]$
σ	Stefan-Boltzmann constant, $[\mathrm{W\,m^{-2}\,K^{-4}}]$

Superscripts

$*$	non-dimensionalized variable

Subscripts

abs	absorbed
an	analytical solution
ap	aperture
$best$	best fit
cav	cavity

ch TMR chamber

$cond$ conduction

$conv$, c convection

$crit$ critical

err error

exp experiment

fc feeding cone

fit data fit

fl fluid line

in incoming, inlet

∞ ambient

ins insulation

ir irradiation

$mean$ mean value

mt mass transport

nom nominal

o outcoming, outlet

opt optical

ow outer wall

p measurement period, particle

$peak$ peak in TMR measurement

$prot$ prototype

pw particle to wall

rad radiation

ran random

$refl$ reflection

ret retention time

rev revolution

sim simulation

sys systematic

th thermal

tot total

var varied

$wall$, w cavity wall

Acronyms

CentRec Centrifugal Particle Receiver

CFD Computational Fluid Dynamic

CNRS Centre National de la Recherche Scientifique

CSP Concentrating Solar Power

DAR Direct Absorption Receiver

DLR German Aerospace Center

FE Finite Element

HFSS	High Flux Solar Simulator
IR	Infrared
LEC	Levelized Electricity Cost
SDOM	Standard Deviation of the Mean
SNL	Sandia National Laboratories
SD	Standard Deviation
TC	Thermocouple
TMR	Temperature Measurement Ring

Chapter 1

Introduction

In times of a growing world population, globalization, scarce resources and increasing environmental pollution, renewable energy technologies are an ecological and sustainable alternative for power generation. Concentrating Solar Power (CSP) systems are considered to be one efficient way of converting solar energy into electricity. In a power tower system, direct solar radiation is concentrated and reflected by mirrors, so-called heliostats, onto a receiver system, where it is collected by a heat transfer medium. The absorbed heat can be either directly used to drive a power block or stored providing dispatchable energy. Load depending power generation and the ability to decouple insolation from electricity production are the main competitive advantages of CSP over other more fluctuating renewable technologies, such as wind and PV.

One of the key elements of a CSP plant is the receiver system. A significant number of potential receiver technologies have been therefore the focus of extensive research over the last few decades. A comprehensive overview of high-temperature central receiver concepts, such as volumetric air, tubular gas and liquid receivers as well as solid particle receivers, is given by Ho &

Iverson [29]. The development of materials, geometric designs, heat-transfer media and processes leading to a high system durability, maximized solar flux capabilities and absorptance along with minimized heat losses are the unique challenges associated with high-temperature receivers.

Investigations from the 1980s have shown that direct absorption receivers (DAR) have the ability to achieve higher outlet temperatures compared to conventional salt in tube receivers [78]. Higher heat losses due to higher temperatures can be overcompensated by gains from the power block efficiency leading to higher overall system efficiencies. Combined with lower receiver and system costs (e.g. parasitic losses $< 1\,\%$) and the inherent feature of directly storing the heat transfer media, reduced levelized electricity costs (LEC) seem achievable [78], [68]. Beside power generation, a broad range of other applications can be addressed with DAR systems, such as solar fuel or high temperature process heat.

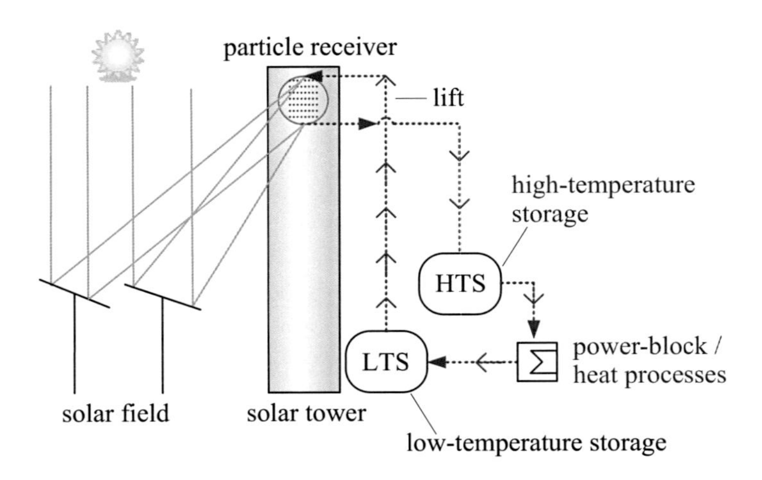

Figure 1.1: Schematic of a particle receiver system with power block and storages.

Within the framework of the development and investigation of DAR systems at the German Aerospace Center (DLR), a solid particle receiver (SPR) concept is developed wih regard to high efficiency, high durability and low cost. A schematic of such a SPR system is shown in Figure 1.1. The main parts of the plant are the receiver, the heat exchanger, a high and low temperature storage, a transport system and the power block, which could be also exchanged by a thermal process or thermo-chemical cycle. Solar radiation is directly absorbed in the receiver by the heat transfer medium which can be either used instantly or be stored in the high-temperature storage. After passing the heat exchanger the cooled down medium is kept in the low-temperature storage for reuse.

1.1 State of the art

Solid particle receivers have been an interest of research for over thirty years. A variety of different particle receivers exists already, but with regard to the receiver concept, which is proposed in the present work, only a selective review of some relevant technologies will be given in the following. The main focus lies hereby on the falling particle receiver and the rotary kiln concept.

1.1.1 Falling particle receiver

With the pioneering work of Martin and Vitko [46] at the Sandia National Laboratories (SNL) in the early 1980s a time of extensive studies on solid particle receivers began. The technical feasibility of a particle receiver was first shown by Hruby [33]. Her investigations focused mainly on the development of a receiver design and the selection of suitable particle materials. Based on the assessment study by Falcone et al. [22], which indicated competitive costs of the solid particle receiver compared to the least expensive air receiver concepts for high temperature applications, the design of a free falling particle receiver was proposed.

A scheme of the concept can be seen in Figure 1.2. The particles are introduced into the receiver at the top and fall through while forming a thin, but optically dense particle curtain. Concentrated sunlight provided by a north field of heliostats enters an inclined aperture and heats the particles directly up to 1000 °C. At the receiver exit the heated particles are collected and transported to the high temperature storage or sent through a heat exchanger providing heat process input. In order to achieve high outlet temperatures and sustain them under part load conditions a recirculation of the particles is provided.

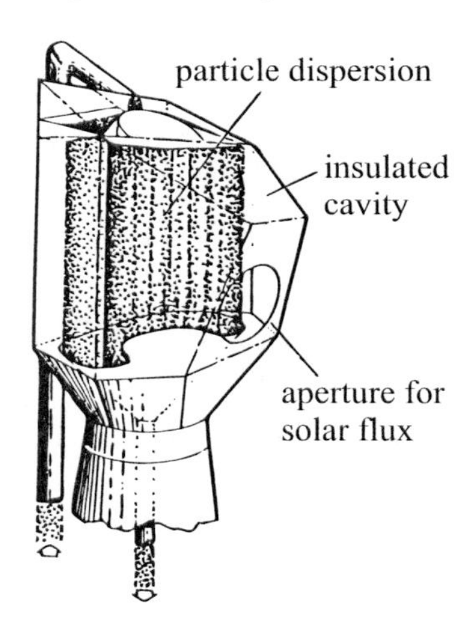

Figure 1.2: Conceptual design of the SNL solid particle receiver [33].

More work regarding flow characteristics and convective heat transfer in a falling particle curtain was studied experimentally and numerically by Hruby et al. [34]. Griffin et al. [26] conducted an extensive evaluation of suitable materials with special attention on optical properties.

Recently, SNL tested a prototype solid particle receiver on-sun [66], providing an experimental basis for the validation of numerical simulations. A computational fluid dynamic (CFD) model of the prototype has been developed by Ho et al. [30] which considered irradiation from the concentrated solar flux, two-band re-radiation and emission from the cavity, discrete-phase particle transport and heat transfer, gas-phase convection, wall conduction as well as radiative and convective heat losses. Comparison with experimental data revealed good agreement with an average relative error of less than 10 %.

A comprehensive overview of studies regarding solid particle receivers is provided by Tan and Chen [70]. Next to a detailed evaluation of design concepts, advantages and disadvantages the authors present results of investigations on factors that influence the receiver efficiency like geometry, particle size and wind.

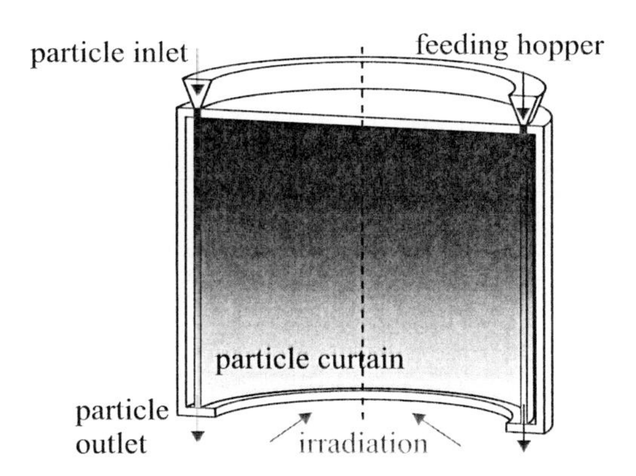

Figure 1.3: Half-section of the falling particle receiver [25].

The DLR currently investigates a similar particle receiver concept, the so-called falling particle receiver [25]. Figure 1.3 shows a sectional view of a possible receiver design. The particles are transported to a feed hopper at the top of a face-down cylindrical cavity. While the particles fall down through an inlet slit to a collecting ring at the receiver bottom, a free falling curtain parallel to the inner cavity wall is formed. Solar radiation enters the receiver through the open aperture and is directly absorbed by the particle curtain. Due to the face-down configuration receiver convection and radiation losses are reduced while the particle curtain is protected against wind influences. An efficient recirculation strategy is employed in order to realize optically dense curtains for high absorptivity and hence increased receiver efficiency.

The falling particle receiver is a promising concept for a highly efficient direct absorption receiver combined with a relative simple design. Preliminary studies by Röger et al. [59] have shown that for a design power of 395 MW_{th} and an outlet temperature of 800 °C a receiver efficiency of about 90 % can be expected. Applying a favorable operation strategy a receiver efficiency of up to 67 % at 20 % part load condition seems possible.

1.1.2 Rotary kiln

The first patent of a rotary solar kiln was filed by Trombe in the late 1950s [74]. He developed a device consisting of a rotating cavity which is directly heated by solar radiation for the application of melting refractory mixtures and different chemical processes. A sectional view of the kiln used for melting is exemplarily sketched in Figure 1.4. The operating rotation rate is sufficiently high that the treated powdery substances are pressed against the wall by centrifugal forces. Depending on the application the speed of rotation may vary from some hundreds to some thousands of revolutions per minute.

Kelbert and Royere [36] from the Centre National de la Recherche Scientifique (CNRS) studied experimentally the thermal performance of a rolling bed rotary kiln at the solar furnace in Odeillo. Ordinary sand with particle sizes from 400 to 600 µm is used as heat carrier. With flux densities of 3 $MW\,m^{-2}$ at the aperture receiver efficiencies from 65 to 91 % at outlet temperatures from 860 up to 940 °C were achieved. The overall efficiency of a whole solar thermal loop including rotary kiln, a hot and cold storage as well as a fluidized bed heat exchanger is evaluated to 65 to 73 % at an inlet power from 50 to 75 kW.

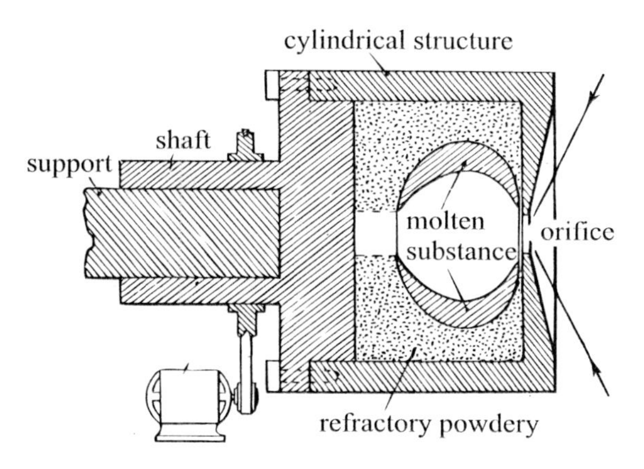

Figure 1.4: Sectional view of a rotary furnace proposed by Trombe [74].

Two reactor concepts for the solar thermal decomposition of zinc oxide were developed at the Paul Scherer Institute (PSI) by Haueter et al. [27] and Müller et al. [50]. Both reactors basically consist of a rotating cavity receiver that is closed to air. Solar radiation is led in through a window which is incorporated in the aperture. ZnO particles are continuously fed into the receiver and forced against the wall due to centrifugal forces. A thick layer of ZnO is formed that insulates and reduces the thermal load on the cavity walls. Tests of the prototypes at the PSI solar furnace have demonstrated the capability of the receivers to be heated up to nearly 2000 K.

For the solar thermal production of lime Meier et al. [48] first developed a 10 kW_{th} rotary kiln with a conical reaction chamber.

The preferred particle flow regime is the continuous rolling motion of the 1 to 5 mm lime particles. First experiments at the solar furnace of the PSI have shown the technical feasibility of the proposed concept.

As an improved design of their former reactor design Meier et al. introduced a $10\,kW_{th}$ multi-tube rotary kiln prototype [47] for the industrial production of lime. In contrast to their previous concept the lime particles are indirectly heated within the absorber tubes. The difficulty in the direct heating technology, where fine calcined particles form a white powder cloud that reflects and absorbs a significant amount of incident radiation, could be therefore avoided. A reliable operation time of over 100 h at solar flux inputs of about $2\,MW\,m^{-2}$ and temperatures up to 1400 K were achieved.

Funken et al. conducted experiments with a rotary kiln reactor for recycling processes of hazardous wastes [23] and aluminium remelting [24] at the DLR solar furnace. Concentrated sunlight is led in through an open radiation duct whose size is optimised regarding minimal radiation losses. Maximum temperatures up to 950 °C at a inlet power level of 3.2 kW were observed.

1.2 The Centrifugal Particle Receiver

One of the major difficulties of the receiver concepts exhibiting a falling particle film is the controllability of the particle retention time in the radiation focus. Espescially for smaller scale power plants where the focus is quite small, recirculation of the particles can be a cost intensive factor. Rotary kilns have been proven to be suitable for high temperature applications as well, although they are mainly developed for chemical processes so far.

In the present work, the idea of the rotary kiln concepts is utilized to develop a particle receiver that exploits centrifugal forces to gain enhanced controllability of the particle behavior and consequently the receiver performance. Due to high working temperatures and a simple and robust design of the receiver, reduced costs of the overall CSP system are expected independent on the plant size. The concept and the functionality of the Centrifugal Particle Receiver (CentRec) is introduced in the following.

1.2.1 Concept

Figure 1.5 presents a schematic of CentRec that basically consists of a fast rotating cylindrical cavity which can be inclined 10 to 90° to the vertical. Small bauxite particles of about 1 mm diameter are used as the heat transfer medium due to their favorable optical properties [26]. The absorption coefficient lies within 0.8 to 0.9 and the specific heat capacity is about $1100\,\mathrm{J\,kg^{-1}\,K^{-1}}$ for the considered temperature range.

The particles enter the cavity through a double-walled feeding cone where they are accelerated to the desired speed. Inside the cavity they are forced against the inner wall due to centrifugal forces and form a thin but optically dense layer covering the entire circumference. Gravitation impels the particles to slowly move downwards in axial direction while they are gradually heated by direct solar radiation that enters the open aperture. The heated particles are gathered by a collector and then either led to the storage or to the associated process (not sketched in the figure). As the receiver can be flexibly tilted in a wide range of angles an optimal inclination angle α regarding convection losses and heliostat field efficiency can be chosen. α lies within 0 and 90°, whereas $\alpha = 0°$ refers to a horizontally aligned receiver with a vertical aperture and $\alpha = 90°$ to a face-down receiver with a horizontal aperture.

In order to achieve a constant outlet temperature at different load conditions an appropriate mass flow rate can be set while adjusting the particle retention time in the receiver by regula-

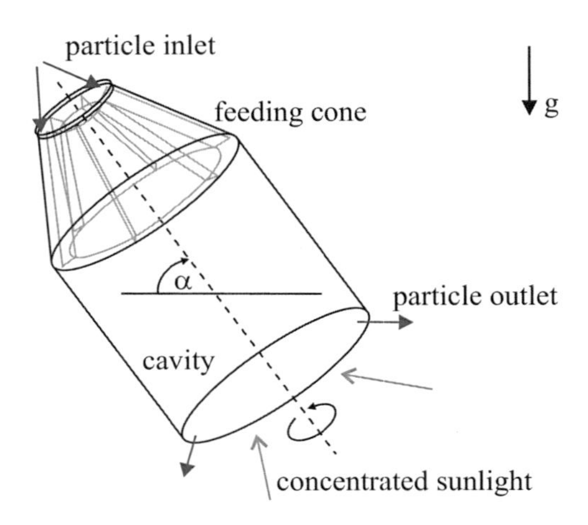

Figure 1.5: Principle design of the Centrifugal Particle Receiver.

ting its rotation speed. As irradiation is directly absorbed by the particles no overtemperature of the receiver materials is existent. Moreover, high peak flux regions are avoided due to the homogenizing effect of receiver rotation. High part load efficiencies are therefore expected.

1.2.2 Receiver dimensioning

The performance of a solar power plant is strongly dependent on its heliostat field efficiency and the receiver efficiency. For minimized convection losses and therefore maximum receiver efficiency a face-down configuration ($\alpha = 90°$) with a circular aperture would be a favourable angle. However, for best field efficiencies an ideal inclination angle can be calculated according to selected cost assumptions for field, tower, receiver and O&M. Thus, an optimized receiver inclination angle exists with the highest combined efficiency of the field and the receiver which will lie in between these two values.

To find the optimum α and an appropriate aperture diameter of the receiver annual performance calculations for a plant size of 1 MW_{th} thermal receiver power at design point are done using the DLR heliostat field layout tool HFLCAL [63]. The methodology, employed field parameters and assumptions as well as a detailed list of the results are found in Wu et al. [82]. A simplified receiver model is implemented in the calculations where only constant thermal radiation losses corresponding to an inlet temperature of 600 °C and an outlet temperature of 1000 °C are considered. Convection, conduction and optical losses are neglected.

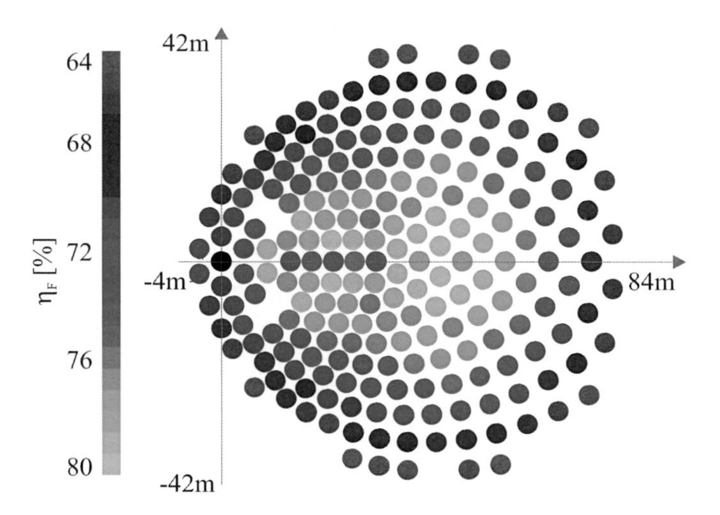

Figure 1.6: Heliostat field layout and efficiency optimized with HFLCAL. 1 MW_{th}, $\alpha = 40°$

Figure 1.6 displays a cost optimized heliostat field for an exemplary 1 MW_{th} power plant in Seville, Spain. The legend on the left indicates the annual heliostat efficiency including cosine losses, blocking and shading, atmospheric extinction and intercept losses. The calculation yields an optimum inclination angle of 40° for which the annual field efficiency lies at 64.20 %.

The field of a size of about 84 m x 88 m contains 215 heliostats with a surface area of $8.24\,\mathrm{m}^2$ each. They are basically installed in a northfield configuration. The tower is 31 m high and the optimized aperture diameter is calculated to be 1.2 m.

Based on the previously defined aperture diameter optical losses for different cavity dimensions are estimated using the DLR ray-tracing tool SPRAY [13]. Various cavity to aperture diameter (D/D_{ap}) and length to diameter ratios (L/D) are studied for two different particle reflectivities of $a = 0.8$ and $a = 0.9$. As indicated in Figure 1.7 the relative reflection loss $d\dot{Q}_{refl}$ diminishes with increasing L/D. Moreover, for higher absorptivities $d\dot{Q}_{refl}$ is decreased but it is enhanced for smaller D/D_{ap}.

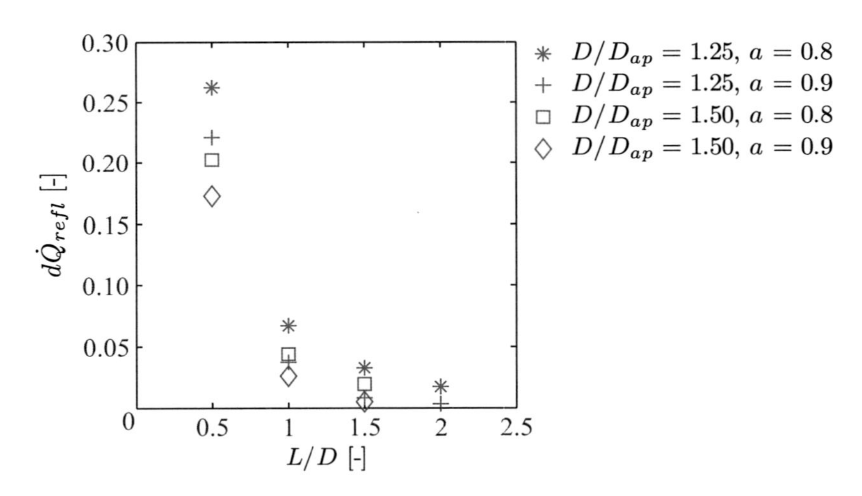

Figure 1.7: Reflection loss calculations for different cavity dimensions using SPRAY [13]. $\dot{q}_{in} = 1\,\mathrm{MW\,m^{-2}}$, $D_{ap} = 1.2\,\mathrm{m}$

As the CentRec concept is considered to be also suitable for bigger power plants of several hundred megawatts a compromise between overall receiver size and reflection losses must be found. Therefore, cavity dimensions of $L/D = 1.5$ and $D/D_{ap} = 1.25$

with optical losses of about 3 % for the conservative case where $a = 0.8$ are the preferred choice for the subsequent receiver design.

1.3 Objective

The objective of the present work is the proof of concept of the proposed receiver design and the demonstration of its feasibility for the desired applications. A prototype in laboratory scale is therefore developed and investigated. As particle motion is strongly dependent on the interaction of gravitational and centrifugal forces, extensive studies regarding the particle behavior and espescially its controllability are conducted. Subsequently, the prototype is tested at the DLR High-Flux Solar Simulator on the one hand to evaluate the overall thermal receiver performances and on the other hand to provide a data base for the validation of the numerical receiver model. Experiments and model provide a comprehensive set of data that aids the determination and understanding of important factors influencing the receiver performance.

Chapter 2

Convection in Rotating Cavities

In order to determine the thermal receiver efficiency of the proposed concept, three major heat transfer mechanisms need to be considered: radiation, conduction and convection. The first two mechanisms can be measured experimentally and estimated theoretically with manageable effort. The determination of the latter one however is quite complex due to the coupling of the temperature and the velocity of the working fluid. Moreover, it strongly depends on other factors such as external wind conditions, receiver wall temperature as well as receiver geometry and inclination.

Despite the comprehensive research regarding convective losses in solar cavity receivers and convection in rotating systems, a combination of both has not been studied so far. Most of the investigations, taking rotation into account, deal with cylindrical enclosures with heated end walls (Rayleigh-Bénard cell). For the present work however, an one end open cylinder with heated side walls needs to be considered as an opening is necessary for the solar energy to enter the receiver (see Figure 1.5). Due to the strong influence of bounding walls and the complex interaction between buoyancy and centrifugal forces on the convective flow, experimental investigations are conducted in order to evaluate the effect of rotation on convective losses for the special case of the CentRec concept. The set-up of the test rig, including instrumentation and experimental procedure as well as the main findings are presented in the following sections.

2.1 State of the art

In this section, an overview is given about substantial investigations of the past few decades, with focus on studies regarding convection losses in solar cavity receivers and convection in rotating systems.

2.1.1 Convection losses in solar cavity receivers

Much research has been done regarding convection losses in stationary cavity receivers. Le Quere et al. [40] first proposed a simple Nusselt number correlation based on his experimental studies of an open cubical cavity with isothermal walls, accounting for the effects of varying receiver temperature and inclination angle. An implicit model describing the convection loss mechanism under no wind conditions was developed in detail by Clausing [17, 18]. Based on his assumptions, he proposed an analytical model for large cavity receiers with Rayleigh numbers ranging from 3×10^7 to 3×10^{10}, taking geometry and tilt angle variations into account. Comparisons with experimental data showed good agreement.

Siebers and Kraabel [64] performed experimental studies on a large 2x2 m cavity with a design surface temperature of up to 800 °C. Their proposed correlation considers the physical property variations in strongly heated cavities.

The convective fluid flow inside a cavity receiver was made visible by Yeh et al. [84] in their experimental investigations of a model cavity using water as working fluid. They presented photographs of convective flow, which was indicated by blue dyed water streamlines, for different Grashof numbers and inclination angles. Similar flow patterns to the predictions by Clausing were found and a Grashof number of 10^7 was identified where flow transition from laminar to turbulent takes place.

A series of additional experimental (e.g. [28], [16], [60], [15], [14]) and numerical (e.g. [56], [57], [10], [61]) investigations regarding convection losses in cubical and rectangular cavities have been carried out. A comprehensive literature review is given in Wu et al. [80].

A Nusselt number correlation for convection losses in cylindrical cavity receivers was first proposed by Koenig and Marvin [38]. Further models were developed accounting for aperture size, surface temperature, tilt angle [69] and various geometries [41]. Paitoonsurikarn et al. [52, 53, 54, 55] conducted numerous experimental and numerical investigations leading to the development of a series of additional correlations.

An electrically heated model cavity receiver was studied experimentally and numerically by Taumoefolau et al. [71] for various tilt angles and surface temperatures from 450 to 650 °C. Comparing their results to ones reported in the literature revealed best agreement with Clausing's model.

Experimental and numerical investigations were carried out by Prakash et al. [58] to accurately determine the stagnation and convective zones in a cylindrical cavity receiver. They proposed a quantitative estimate to identify the zone boundary in terms of a so-called "critical air temperature gradient". Locations with air temperature gradients less than this critical value represent the stagnation zone while locations with gradients exceeding this value built the convective zone.

Most recently, Wu et. al reported numerical [81] and experimental [79] studies, investigating the effect of aperture characteristics, like position and size, and the influence of differentially heated bounding walls, respectively. Empirical correlations have been proposed relating convection, radiation and total heat loss Nusselt numbers to the Grashof number, tilt angle and ambient temperature.

2.1.2 Rotating convection

Especially in science fields like geology, oceanogrophy, climatology or astrophysics lots of intensive research regarding rotating convection have been done. Boubnov & Golitsyn [12] e.g. did extensive studies regarding convection in rotating fluids. They experimentally observed a ring pattern of convection flow resulting from the fluid spin-up and vertex interactions between two adjacent vortices.

Homsy & Hudson [31, 32] presented an analytical approach of centrifugally driven thermal convection in a vertical rotating cylinder heated from below. Top and bottom ends exhibit constant but different temperatures. As the solution strongly depends on the boundary condition, they used boundary layer methods to obtain solutions on top, bottom and the inviscid core of the cylinder for various boundary conditions, such as insulated or perfectly conducting side walls.

Experiments were conducted by Yang et al. [83], who studied natural convection inside a closed horizontal cylinder with differentially-heated ends for the application of crystal growth. They investigated the effect of rotation on the heat transfer for a constant Grashof number at 1.43×10^6 and air as the working fluid. They found out that at small rotation rates the heat flux distribution is rendered more uniform. With increasing rotation the buoyancy driving force becomes more and more suppressed resulting in lower heat transfer rates. At sufficiently high rotation speeds the flow acts like a rigid body and the main heat transfer is by conduction only.

Experimental studies of rotating convection in a vertical cylinder heated from below by Ker et al. [37] revealed the stabilizing effect of rotation for certain rotation rates. At a given temperature gradient two different ranges of rotation rates exist where the flow is stabilized. Outside these ranges it oscillates periodically or quasi-periodically in time.

Lin et al. [43] conducted experiments in an air-filled heated inclined cylinder rotating about its vertical axis. Time resolved temperature measurements were recorded indicating stabilized and destabilized flow under different conditions.

Recent experiments and direct numerical simulations of turbulent rotating Rayleigh-Benard convection [39, 85] underlines the previous experimental findings. It was shown that heat transfer under certain conditions is enhanced despite of the stabilizing effect of rotation. At modest rotation rates, the so-called Ekman-pumping, embodied by an array of strongly localised vertical vortices, is considered to be responsible for the vertical transport of fluid and heat. However, with increasing rotation rate the stabilizing effect of rotation dominates the flow and the heat transfer is decreased again.

2.2 Theory

Natural convection in rotating fields results in very complex transport phenomena in the fluid. In order to determine the actual fluid motion centrifugal and Coriolis forces need to be taken into account next to buoyancy, momentum and viscous effects. A brief summary of the underlaying theory as well as characteristic dimensionless parameters regarding convection losses in cavity receivers is therefore given in this section.

2.2.1 Convective heat loss mechanisms in cavity receivers

In order to briefly describe the basic convective heat loss mechanisms in cavity receivers, the theory proposed by Clausing [17] is introduced. For the simple case of natural convection, when no wind effects are present, the cavity interior can be divided into two regions: a convective and a stagnant zone. Most of the fluid motion due to temperature gradients between hot cavity walls and cold ambient air happens in the convective zone, where mass and heat is transfered across the aperture.

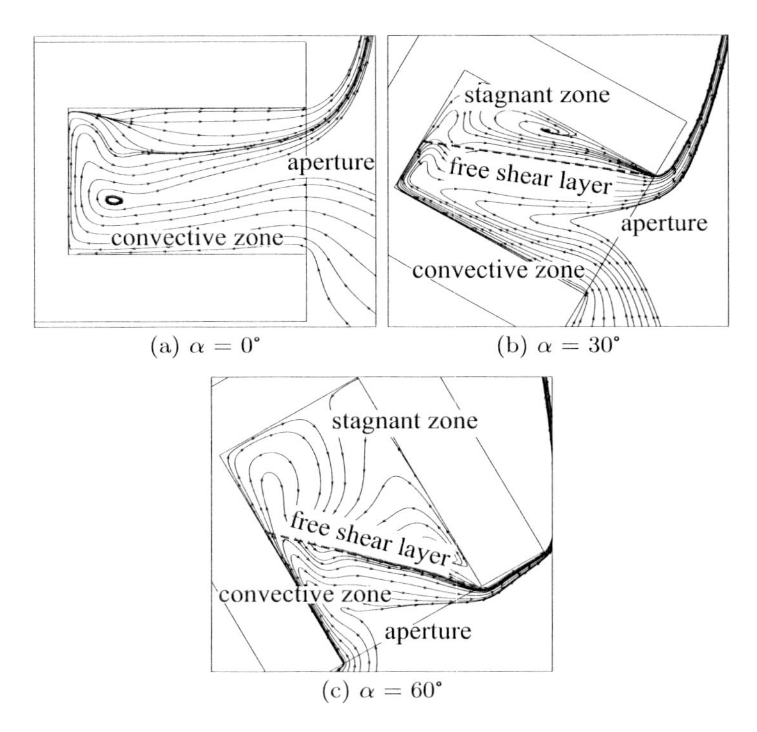

(a) $\alpha = 0°$ (b) $\alpha = 30°$ (c) $\alpha = 60°$

Figure 2.1: Streamlines of convective flow in a cavity receiver at three different inclination angles α.

Typical flow patterns in a cavity receiver are shown in Figure 2.1. When the receiver is oriented horizontally with a vertical aperture area (Figure 2.1a), no stagnant zone can be noticed. The entire cavity is filled by the convective zone, where cold ambient air enters the cavity at the lower half of the aperture and is then heated up at the lower hot wall. When encountering the back wall the air flows upwards, reverses along the upper wall and exits the cavity as a hot plume. However, increasing the receiver inclination angle, with the aperture facing downwards, the stagnant zone becomes dominant resulting in a stable stratification of the flow above the free shear layer (Figure 2.1b, 2.1c). This layer, seperating the two different zones, moves closer to the aperture as the inclination angle increases, indicating the expanding dominance of the stagnant zone. Diminishing convective losses are the consequence culminating in negligible small losses for an inclination angle of 90°.

2.2.2 Characteristic dimensionless parameters

Four independent dimensionless parameters are introduced due to their importance for the present system. The Prandtl number

$$Pr = \frac{\nu}{\kappa}. \tag{2.1}$$

characterizes the dissipative properties of the fluid and therefore defines the relation between kinematic viscosity ν and thermal diffusivity κ. In order to describe the respective relevance of buoyancy and dissipation in the fluid, the Rayleigh number is used,

$$Ra = \frac{g\beta\Delta T L^3}{\nu\kappa}, \tag{2.2}$$

where β denotes the thermal expansion coefficient, g the gravitational constant, T the temperature and L a reference length. A further important parameter is the Taylor number

$$Ta = \frac{4\Omega^2 L^4}{\nu^2}, \tag{2.3}$$

which relates the centrifugal and viscous forces with Ω as the angular rotation speed. The respective importance of buoyancy and rotation can be related by the dimensionless Rossby number which is formed as [35]

$$Ro = \frac{\sqrt{g\beta\Delta T/L}}{2\Omega} = \sqrt{\frac{Ra}{Pr\,Ta}}. \tag{2.4}$$

Finally, the Nusselt number describes the relation between the thermal convection and thermal diffusive conduction,

$$Nu = \frac{hL}{\lambda}, \tag{2.5}$$

with h as the heat transfer coefficient and λ as the thermal conductivity.

2.3 Experimental set-up and procedure

In order to measure the convection losses experimentally under laboratory conditions, an electrically heated rotating cavity has been designed and built up. The main features and measurement techniques of the set-up are subsequently described. A total of 90 experiments have been conducted, consisting of measurements of six different rotation speeds (0 rpm, 30 rpm, 60 rpm, 90 rpm, 120 rpm and 150 rpm) at three mean wall temperatures (400 °C, 600 °C and 800 °C) and five inclination angles (0°, 20°, 40°, 60° and 80°).

2.3.1 Test receiver

A cross-sectional view of the model cavity is presented in Figure 2.2. The inner cylinder basically consists of a 5 mm thick, high-temperature steel tube with an inner diameter of 158 mm and 245 mm length. Since the investigations solely focus on thermal losses, three heating coils, with a maximum power of 2300 W each, are applied and spirally wound around the outer cavity shell providing the desired heat input. As the influence of the back wall is of minor importance, it is not heated and consists of a simple 2 mm thick steel plate. Several layers of insulation material are used preventing excessive conduction losses to the ambient.

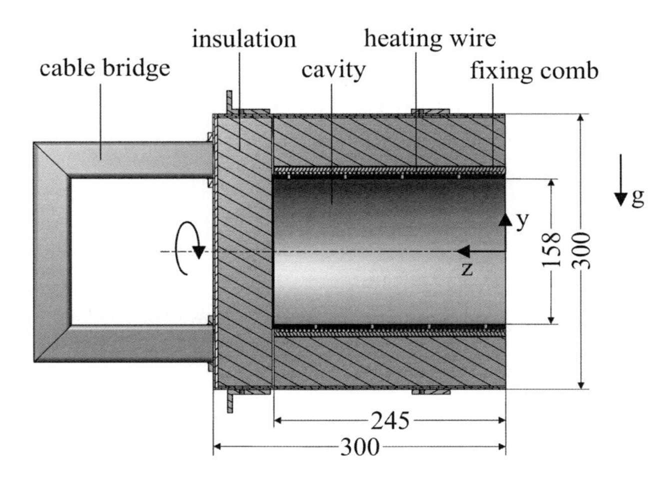

Figure 2.2: Cross section of model cavity. Dimensions are in mm.

For the precise measurement of radiation losses the entire inner cavity surface is coated with high-temperature resistant black paint with a defined emission coefficient [45]. The entire insulated cavity is mounted in a steel cylinder (diameter and length of 300 mm) which is pivoted in a support structure as shown in Figure 2.3. The cavity can be rotated about the x-axis by different inclination angles. For the rotation about its axial axis (z) , an electrical DC motor is used actuating the cavity via a drive wheel. A description of the whole test rig is detailed in Waibel [76].

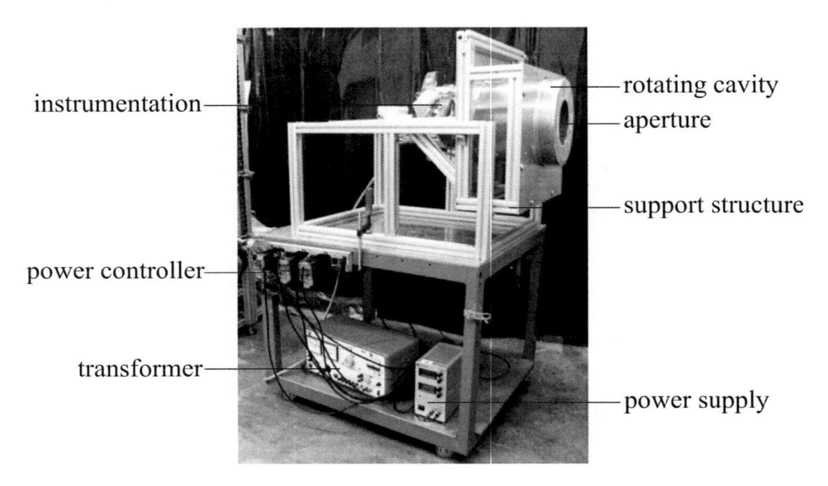

Figure 2.3: Experimental setup in the laboratory.

2.3.2 Instrumentation

In order to determine the convection losses, two important quantities are measured: receiver wall temperature and consumed heating power, which corresponds to the overall thermal losses. Knowing conduction ($\dot{Q}_{cond}$) and radiation losses ($\dot{Q}_{rad}$), convection losses can be simply calculated by substracting them from the electrical input power P, such that

$$\dot{Q}_{conv} = P - \dot{Q}_{rad} - \dot{Q}_{cond}. \tag{2.6}$$

The schematic diagramm in Figure 2.4 describes the measurement instrumentation. The electrical heating is controlled by three power controllers, where the voltage of the heating cables is set by phase cutting. As the electrical resistance of the cables changes with the temperature, the actuall heating power needs to be monitored by three measuring transducers for active power. Power controllers and measuring transducers are regulated and read out by a computer via a data acquisition board.

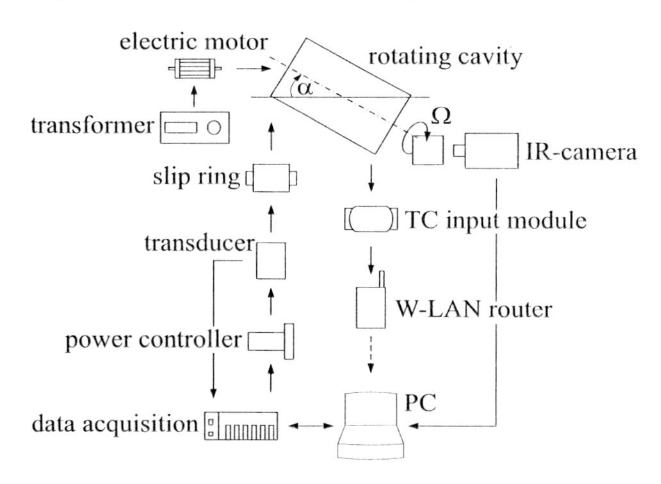

Figure 2.4: Instrumentation set-up of the convection test rig [76].

The mean cavity surface temperature is determined by 21 type N thermocouples (TC). 16 are placed on the cylindrical cavity side wall within the wall thickness with four times four TCs in a row. Five TCs are used to measure the temperature at the back wall, arranged concentrically around the center in a distance of 17 mm from each other. A sketch of the thermocouple distribution is presented in Figure 2.5.

The TCs are connected to three thermocouple input modules, mounted on the external surface of the steel cylinder. Since the whole cylinder is rotating, the modules are linked to a wireless LAN router that transmits the received temperature data via radio. The energy supply of heating wires, modules as well as the router is realized through a slip ring, installed at the end of

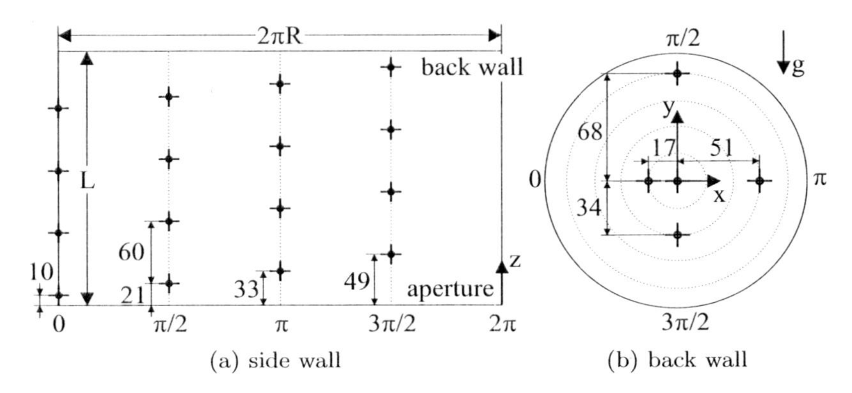

Figure 2.5: Positions of thermocouples in the cavity side and back wall. Distance declarations are in mm.

the cable brigde. A thermographic camera is used in addition to observe the temperature distribution of the cavity inner surface. The DC motor driving the cavity is manually set by a power transformer whereas the rotation speed is measured by a digital tachometer.

As measuring programm a LabVIEW routine is used recording the temperature data and controlling the heating power. Every single thermocouple is read out with a sampling rate of 1 Hz while the active power measured by the transducers is recorded with a sampling rate of 10 Hz.

2.3.3 Experimental procedure

In order to prevent critical thermal stresses during the heating process of the model cavity, high temperature gradients are avoided by gradually increasing the electrical power. Steady state is reached after about 2 hours when the variation of the mean wall temperature is less than $0.1\,\mathrm{K\,min^{-1}}$. As the power controllers of the heating coils exhibit slight shifts in the provided power level, they are regulated by a PID controller to keep the mean surface temperature constant. During one experimental cycle, inclination angle and temperature level are fixed while the rotation rate is altered. All measured quantities are derived from the mean of a measurement period of 40 minutes.

2.3.4 Measuring conduction losses

Air gaps due to imperfectly built insulation and thermal bridges due to the installation of thermocouples on the cavity walls lead to additional conduction losses, which need to be considered. The thermal conductivity of the test rig can therefore not be described by the thermal conductivity of the insulation material only. Experiments are conducted to evaluate the conduction loss $\dot{Q}_{cond}$ of the overall system by closing the aperture with an insulated plug, preventing radiation and convection losses through the cavity opening. Substracting the heat losses through the plug from the consumed heating power results in the conduction losses of the model cavity at defined receiver temperatures.

Experiments are carried out with a stationary cavity for five different mean wall temperatures in a range from 300 °C to 500 °C showing an expected linear dependence of $\dot{Q}_{cond}$ to temperature (Figure 2.6). The effect of rotation on $\dot{Q}_{cond}$ due to forced convection on the outer receiver surface is examined exemplarily for a rotation speed of 120 rpm. Figure 2.6 indicates a slight increase of conduction loss by 1.4 %, which is within the experimental uncertainty and can be therefore neglected. The influence of inclination angle is also negligible as the thermal conductivity is mainly dependend on material properties and design characteristics, which are not altered by the tilt angle. For determination of the convection losses the linear dependence of $\dot{Q}_{cond}$ is thus utilized by using the correlation, fitted on the measured data, whereas

$$\dot{Q}_{cond,fit} = [0.5428\,(\overline{T}_{w-\infty}/^{\circ}\mathrm{C}) - 78.0613]\ \mathrm{W}, \qquad 25\,^{\circ}\mathrm{C} \leq \overline{T}_{w-\infty} \leq 900\,^{\circ}\mathrm{C} \tag{2.7}$$

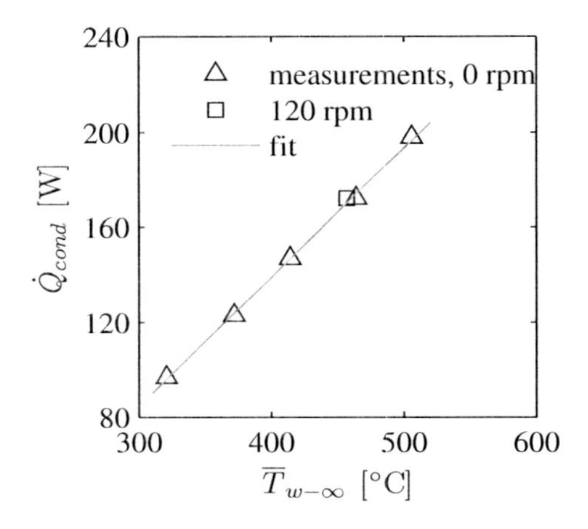

Figure 2.6: Experimental measurement of conduction losses through the cavity wall. $\overline{T}_{w-\infty} = \overline{T}_w - T_\infty$ denotes the difference between the area averaged wall temperature of the cavity surface $\overline{T}_w$ and the ambient temperature T_∞.

2.3.5 Calculating radiation losses

Radiation losses are determined numerically using the surface-to-surface radiation model in Ansys Fluent [8]. The model is used to account for the radiation exchange in an enclosure of gray-diffuse surfaces, where the energy exchange between two surfaces is calculated via their radiosity and view factors.

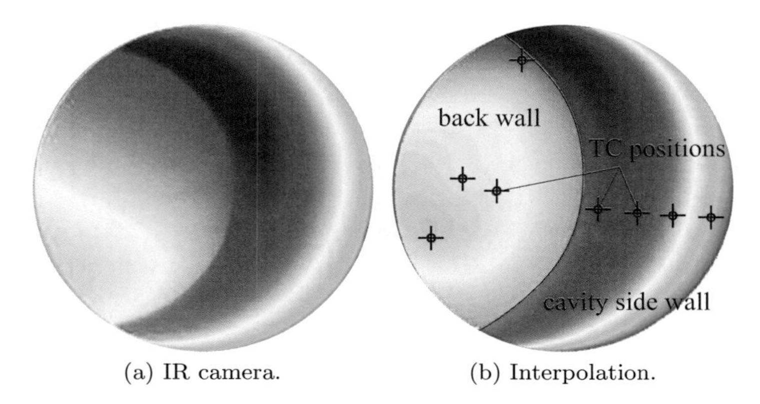

(a) IR camera. (b) Interpolation.

Figure 2.7: Qualitative comparison of interpolated wall temperature distribution to IR camera measurements. The temperature scale is neglected as the recordings are just compared qualitatively.

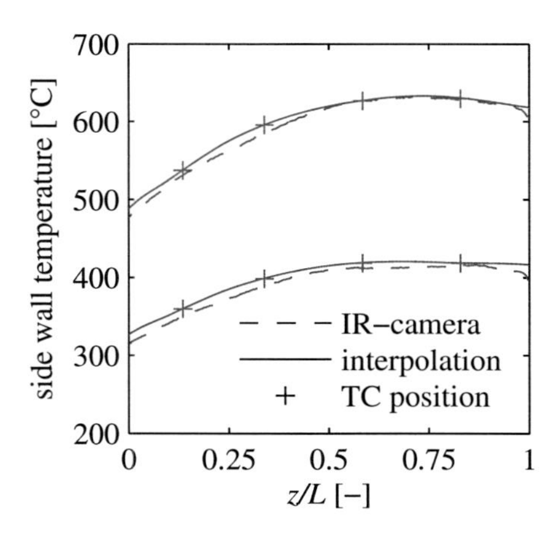

Figure 2.8: Comparison of interpolated wall temperature distribution to measurements by the IR-camera along the dashed line in Figure 2.7b.

As a boundary condition on the cavity surface, an interpolated temperature profile is applied, which is based on the experimentally measured temperatures at the discrete locations. The surface temperature distribution, obtained by a cubical interpolation method, is compared qualitatively to records of the infrared (IR) camera, where good agreement is found (Figure 2.7). Just on the cavity back wall, interpolation results are poor due to the small number of installed thermocouples. As the influence of the back wall plays a minor role, the discrepancies are accepted.

According to Figure 2.8, the quantitative comparison between interpolated and measured temperatures along the dashed line in Figure 2.7b reveals a difference of about maximum 2 %.

2.3.6 Uncertainty analysis

In order to analyse the uncertainty of the determined convection losses, the error propagation after [72] is considered as $\dot{Q}_{conv}$ is not directly measured, but calculated according to (2.6). The absolute uncertainty $\delta\dot{Q}_{conv}$ is therefore defined by the sum of the uncertainties of each term,

$$\delta\dot{Q}_{conv} = \delta P + \delta\dot{Q}_{cond} + \delta\dot{Q}_{rad}. \tag{2.8}$$

The relative uncertainty of the measured electrical input power $\delta P/P$ is about 0.7 %, dictated by the given accuracy of the measuring transducers and the data aquisition module.

Conduction losses are directly determined with the same uncertainty, but as they are estimated by (2.7) for the actual convection loss calculation, a maximum error of 1 % for the fitted correlation is taken. Additionally, when calculating $\dot{Q}_{cond}$ the accuracy of the temperature measurements needs to be accounted for as the correlation is dependent on the mean wall temperature $\overline{T}_w$. The systematic uncertainty of $\delta\overline{T}_w$ consists of errors of about 1.5 %, given by the manufacturer accuracy of the used thermocouples and input modules, and on uncertainties due to the small distance between TC junction and cavity inner wall surface. Statistical errors of $\overline{T}_w$ are defined as the standard deviation of the mean (SDOM) [72]. The SDOM varies depending on the actual temperature distribution with a maximum of about 1.8 %. Measurement uncertainties of the ambient temperature are neglected as the laboratory's temperature was held constantly at $T_\infty = 20\,°C$ by an air conditioner. Combining all uncertainties in a quadratic sum, a total error of about 2.5 % is estimated for the conduction losses.

Radiation losses are determined by radiosity calculation. As the grid resolution of the model is sufficiently fine, numerical uncertainties are negligible. Errors due to the assumption of grey diffuse radiation exchange are also neglected due to the special coating on the receiver walls with an almost wavelength independent emissivity [45]. Moreover, comparisons of simulated with analytically calculated $\dot{Q}_{rad}$ for constant wall temperatures revealed excellent agreement. The uncertainty is therefore only given by the applied boundary condition for the surface temperatures and the emissivity coefficient ε of the wall coating, which lies between 0.87 and 0.90 for the considered temperature range.

It is shown before, that the used interpolation method predicts the actual temperature distribution on the cavity surface with an acceptable accuracy. Quantitative comparisons with infrared recordings exhibit uncertainties of about 2 %, which will be considered as the relative error of the applied temperatures. As temperature is considered with the fourth power in the radiation loss calculation the uncertainty $\delta\dot{Q}_{rad}$ is quite high due to error propagation. Considering the increase of $\delta\dot{Q}_{rad}$ with higher temperatures and accounting for the uncertainty of ε, an maximum error of 9 % is estimated for the case with a mean surface temperature of $\overline{T}_w = 800\,°C$.

By summing all uncertainties accordingly to (2.8) a maximum total error of $\delta\dot{Q}_{conv} = 12.5\,\%$ is expected for the measured convection losses in the present experiments. A more detailed uncertainty analysis is found in Waibel [76].

2.4 Results and discussion

The experimental parameters are summarized in Table 2.1 for the subsequent discussion of the results. *Ra* and *Ro* are calculated according to Clausing [18] using thermophysical properties at film temperature $T_f = (T_w + T_\infty)/2$, and the cavity diameter as a constant reference length for all inclination angles.

Table 2.1: Experimental parameters

$\overline{T}_w$	400 °C	600 °C	800 °C
Ra [-]	1.59×10^7	1.06×10^7	8.48×10^6
Ω [rpm]		*Ro* [-]	
0	∞	∞	∞
30	1.11	1.25	1.34
60	0.56	0.63	0.67
90	0.37	0.42	0.45
120	0.28	0.31	0.34
150	0.22	0.25	0.27

2.4.1 Heat transfer

The dependence of the convective heat losses (*Nu*) on rotation (*Ro*) is shown in Figure 2.9. Results for three different temperatures as well as for various inclination angles are presented. Since the focus lies on the qualitative effect of rotation in this section, *Nu* is normalised by its non-rotating value at $Ro \to \infty$. Error bars are left out as the changes in convective losses due to rotation lie in between the estimated error. However, all experiments were repeated several times and the same qualitative tendency could be measured.

Figure 2.9a clearly reveals the stability effect of rotation at smaller Ro , leading to smaller convective heat losses. For an horizontal orientation of the cavity, rotation decreases convective heat transfer down to $> 10\,\%$. This trend becomes less pronounced with increasing inclination angle and it even reverses for $\alpha \geq 60°$, where heat transfer seems to be enhanced by rotation. Examining Figure 2.9b and 2.9c the same trend of Nu is observed, although it is less distinct for $\alpha \leq 60°$.

The effect of rotation on convective heat losses in cavity receivers is obviously strongly dependend on the cavity orientation. With the aperture facing more downwards, the present configuration approaches the rotating Rayleigh-Bernard cell, investigated by [39, 85], where similar results regarding enhanced heat transfer for certain ranges of Ro have been found. However, the Rayleigh-Bernard cell is an enclosed cylinder with heated bottom while our cavity is one-side open with heated side walls. The enhanced heat transfer can therefore not be explained with the Ekman pumping mechanism. It is assumed instead that the stagnation zone inside the cavity, which increases and moves closer to the aperture with larger α, is disturbed by the mildly swirling flow at moderate rotation rates, leading to higher convective heat losses. By increasing Ω the stabilizing effect of rotation would probably prevail as in the cases for $\alpha < 60°$ and the heat transfer would decrease again. Due to the limited dimensioning of the experimental set-up higher rotation rates of $\Omega > 150\,\mathrm{rpm}$ could not be realized which is why this last statement is not fully proven at this point. The dependence of the relative Nusselt number on the receiver orientation at different rotation rates is shown in Figure 2.10. It is obvious, that α far more influences the convective heat transfer compared to Ω (up to $60\,\%$). As already noticed by Clausing [18], convective heat losses decrease with increasing α

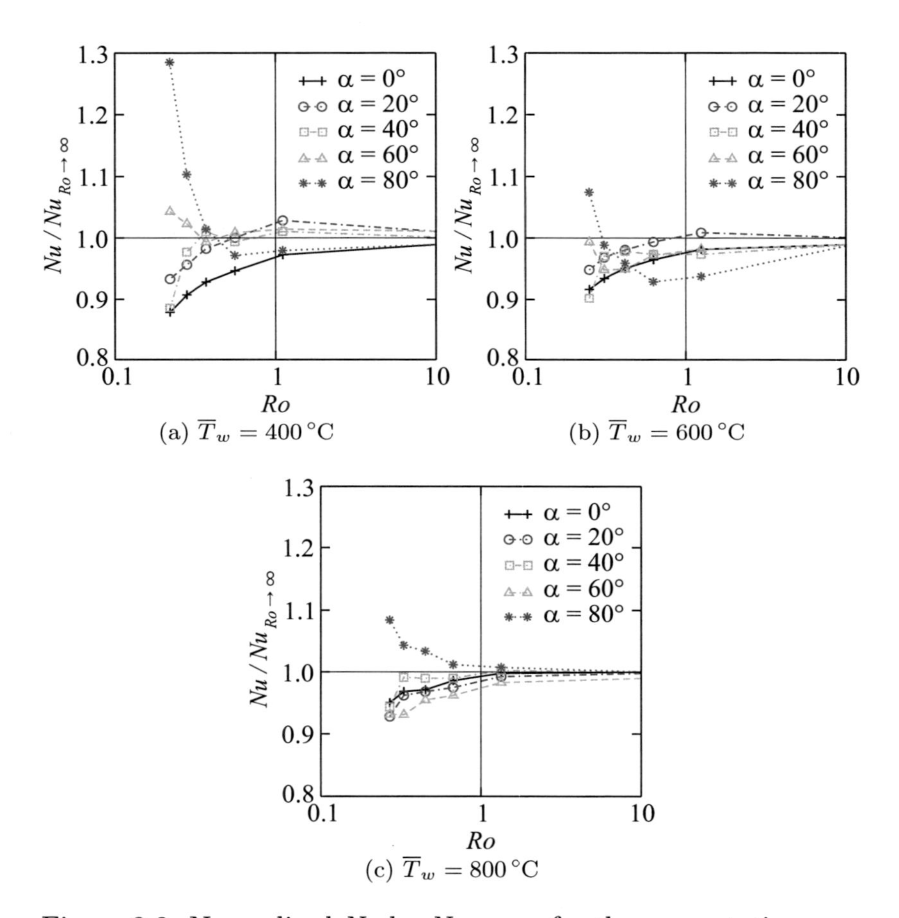

(a) $\overline{T}_w = 400\,°\mathrm{C}$

(b) $\overline{T}_w = 600\,°\mathrm{C}$

(c) $\overline{T}_w = 800\,°\mathrm{C}$

Figure 2.9: Normalised Nu by $Nu_{Ro\to\infty}$ for the non-rotating case as a function of Ro for different $\overline{T}_w$ and α.

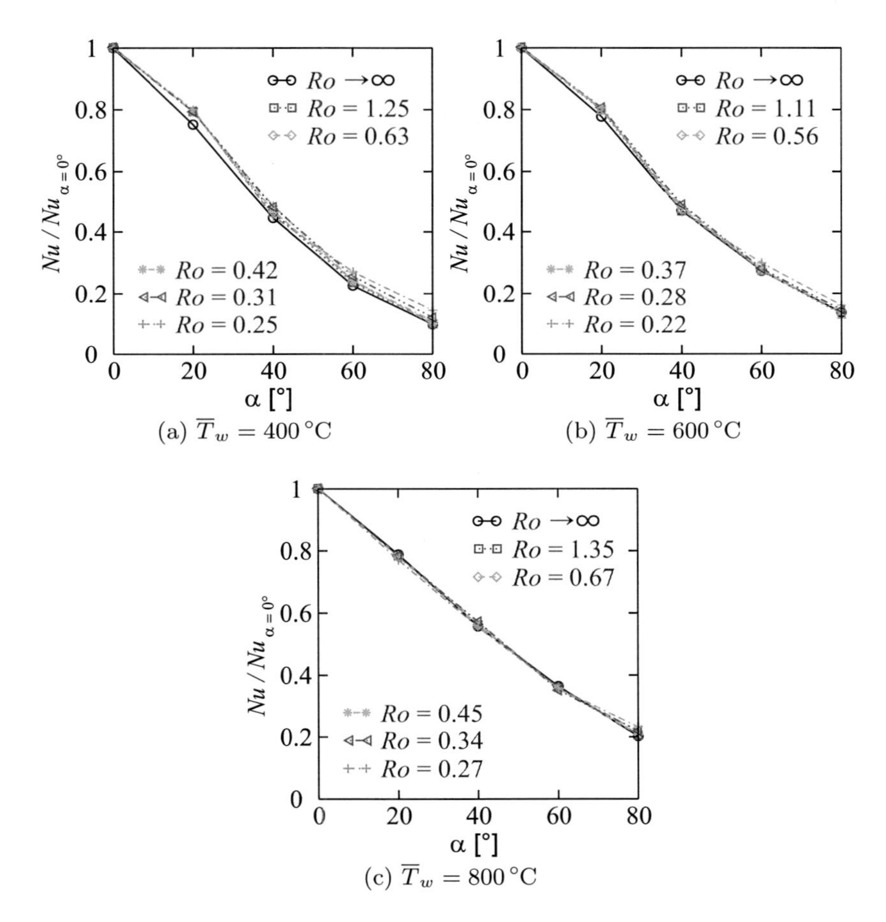

(a) $\overline{T}_w = 400\,°C$

(b) $\overline{T}_w = 600\,°C$

(c) $\overline{T}_w = 800\,°C$

Figure 2.10: Normalised Nu by $Nu_{\alpha=0°}$ for the horizontal case as a function of α for different $\overline{T}_w$ and Ro.

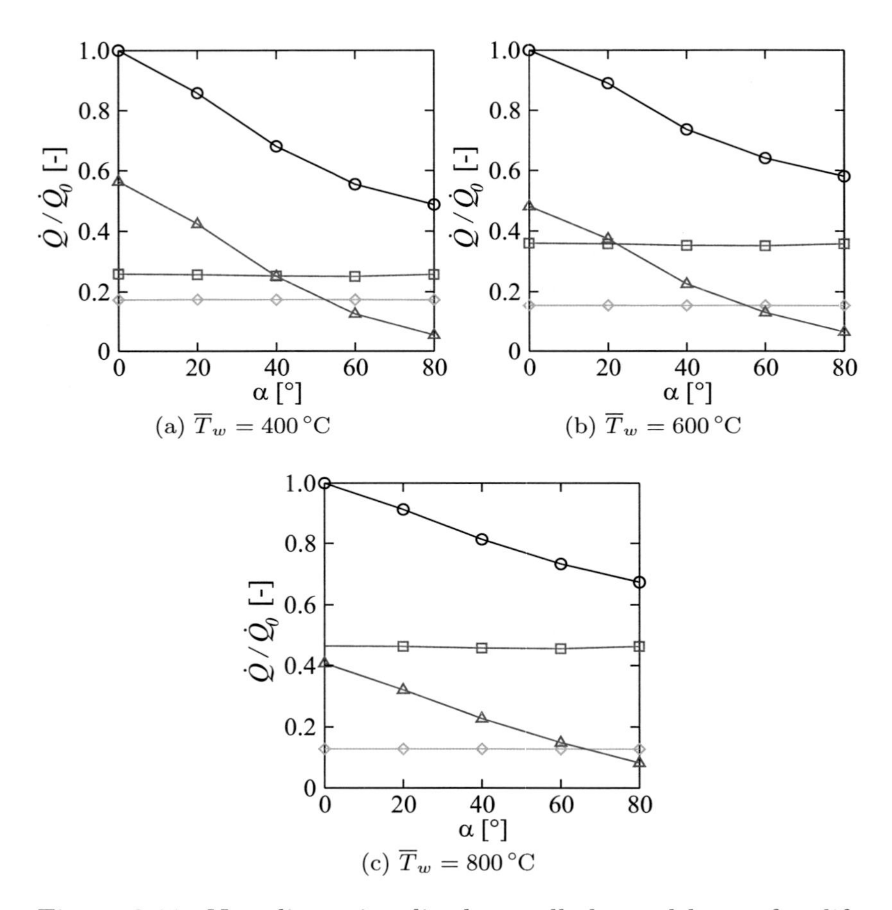

(a) $\overline{T}_w = 400\,°\mathrm{C}$

(b) $\overline{T}_w = 600\,°\mathrm{C}$

(c) $\overline{T}_w = 800\,°\mathrm{C}$

Figure 2.11: Non-dimensionalized overall thermal losses for different temperatures. (○–○) $\dot{Q}_{overall}$, (□–□) $\dot{Q}_{rad}$, (◇–◇) $\dot{Q}_{cond}$, (△–△) $\dot{Q}_{conv}$.

due to the development of the stagnation zone inside the cavity. Considering now the overall thermal loss and its composition in Figure 2.11, the percentage of convective losses is pretty small (about 10 %) for high α. Conduction and radiation losses take precedence due to their independence of the receiver orientation. Moreover, with a maximum of about 66 %, radiation represents the largest share of overall thermal losses as the temperature increases.

For the present application of a solar thermal receiver the quantitative effect of rotation on $\dot{Q}_{conv}$ is of main interest. Looking at Figure 2.9 the convective heat transfer is affected about ± 10 % by rotation in the range of considered Ω, except in the case with $\overline{T}_w = 400\,°\mathrm{C}$ and $\alpha = 80°$, which can be seen as an outlier.

2.4.2 Temperature distribution

The interpolated temperature distribution of the cavity side walls for a mean surface temperature of $\overline{T}_w = 600\,°\mathrm{C}$ and various experimental parameters is displayed in Figure 2.12. For clarity purposes the cylindrical cavity surface is shown uncoiled, whereas $\pi/2$ denotes the cavity "top" and $3\pi/2$ the cavity "bottom" (compare to Figure 2.5).

The typical temperature distribution for a stationary, non-rotating receiver in horizontal position ($\alpha = 0°$) is depicted in Figure 2.12a. Cold ambient air entering at the receiver bottom is heated by the hot surface walls and rises up due to its lower density. An irregular temperature distribution with a high temperature spot at the top and a low temperature valley at the bottom is therefore visible. Figure 2.12b indicates however a more uniform temperature distribution when the receiver is rotated. It is assumed, that rotational forces tend to disturb the air circulation inside the cavity leading to circumferentially homogenized temperatures. As the hot spot on the cavity top is vanished,

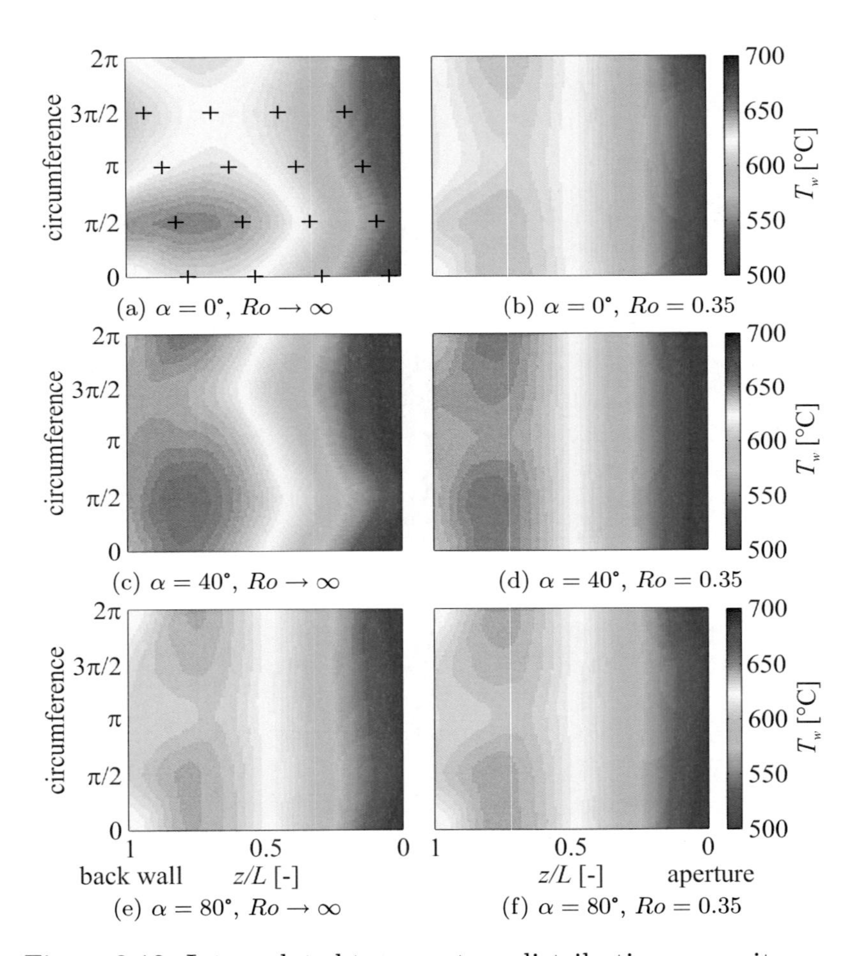

Figure 2.12: Interpolated temperature distribution on cavity surface at different inclination angles α and Ro for $Ra = 2.33 \times 10^8$. For clarity purposes the cavity surface is unwinded. Thermocouple positions, denoted by black crosses, are shown exemplarily in Figure (a).

the temperature gradient is decreased resulting in less convective heat transfer. However, this assumption must be taken with caution. Due to the fast rotating cavity and the thermal inertia of the thermocouples, different temperature zones of the air inside the cavity, as it is present in the non-rotating case, might not be accurately measured. In order to get the final proof of the proposed assumption measurements of the air flow inside the cavity are indispensable.

The influence of the receiver tilt angle is shown on the left of Figure 2.12. As α is increased, with the receiver facing more downwards, the development of a stagnation zone is clearly apparent. Stagnant hot air is “trapped” in this zone resulting in the wall temperature to be distributed more uniform. The homogenized temperature distribution expands with increasing α as clearly indicated in Figure 2.12c and 2.12e. This is thought to be due to the growth of the stagnation zone leading to generally more uniform temperature distribution inside the cavity. The effect of rotation on the wall temperature distribution does not considerably change with increasing α as almost no difference is visible between the figures on the right.

By relating these findings to the results in Figure 2.9, it can be concluded, that a uniform temperature distribution leads to decreased convective heat transfer. However, for the case with $\alpha = 80°$ the effect of rotation can not be derived from the temperature distribution as almost no variation between stationary and rotating receiver is present.

2.4.3 Comparison to literature

The main goal of the current work is to evaluate the thermal efficiency of the introduced receiver concept. Considering the present investigations, the effect of rotation on the overall thermal losses is less than 1 % for the temperature and orientation range the receiver will be operated. Rotational influences can be therefore neglected. On the fluid- and thermodynamics point of view however it is quite interesting to investigate the flow field behavior under rotation in order to illuminate the present findings. Since this is not the main purpose of the present work this matter has to be left to subsequent researchers.

For a simplified model of the overall receiver efficiency, literature correlations seem to be sufficient to estimate convection losses of the receiver. Due to the great variety of different published models a comparison of the present experimental data is done to some selected relevant correlations. In Figure 2.13 the convection losses for the stationary receiver case are compared for varying surface mean temperatures and inclination angles. The best agreeement is found with Clausing's model [18] although it has been developed for higher Rayleigh numbers from 3×10^7 to 3×10^{10}, based on a cubical cavity for large central receiver systems. However, it is a widely generalized correlation as it considers the existence of the convective and stagnant zone inside the cavity depending on the inclination angle and accounts for several important effects, like variable property influences and cavity geometry.

The model from Koenig & Marvin [38] predicts the convection losses quite well in the lower tilt angle regions (for $\alpha \leq 40°$), but underestimates them for higher α. On the contrary, the Paitoonsurikarn & Lovegrove model [55] agrees better for $\alpha \geq 40°$ and underestimates the losses for lower α. Stine & McDonald [69] and Lovegrove et al. [44] under- and overestimate convection losses respectively. All these correlations are derived as a best fit to experimental data and are tailored to the investigated cavity geometries. Almost all of them account for the cavity geometry only by the ratio of aperture to cavity diameter D_{ap}/D_{cav} which is not sufficiently accurate as the cavity depth affects the convective heat transfer as well. Leibfried & Ortjohann [41] considered this fact and extended the Stine & McDonald model for more general

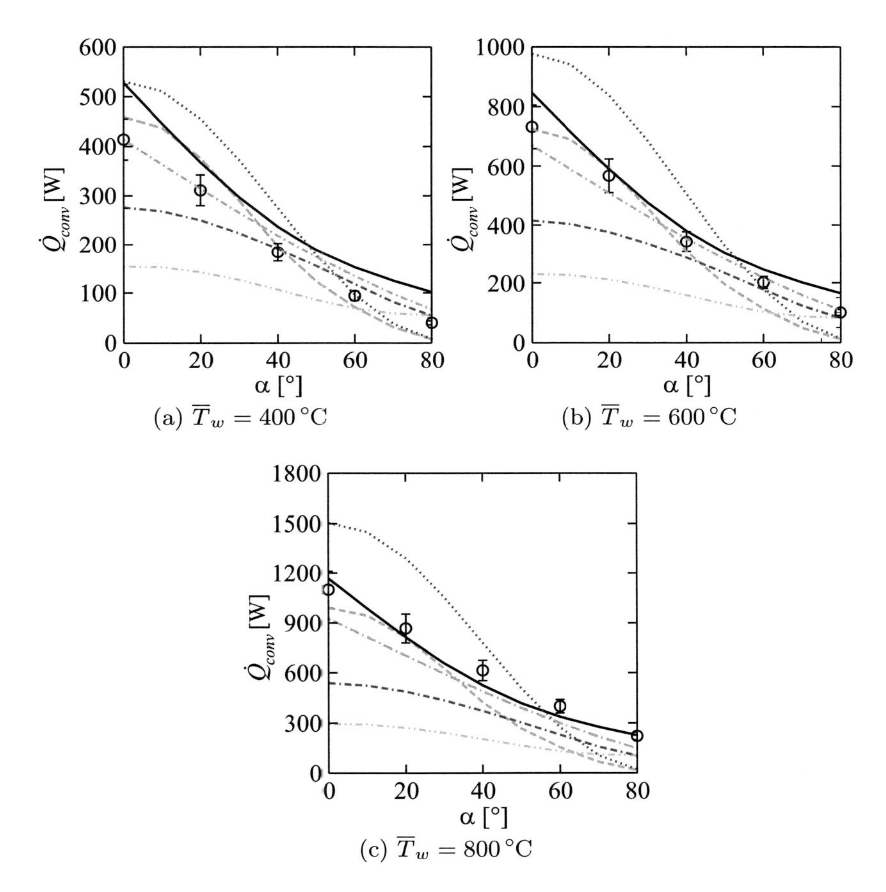

Figure 2.13: Comparison of experimental to literature data for the non-rotating case at different $\overline{T}_w$. (◯) Experimental data including error bars, (- - -) Koenig & Marvin [38], (—) Clausing [18], (· · ·) Stine & McDonald [69], (- · -) Leibfried & Ortjohann [41], (- · ·) Lovegrove et al. [44], (· - -) Paitoonsurikarn & Lovegrove [55]

geometries by taking the ratio of aperture to cavity area A_{ap}/A_{cav} into account. The agreement with present experimental data, especially for $\overline{T}_w = 400\,°C$ and $\overline{T}_w = 600\,°C$, is quite good.

Since Clausing's model suits the best for higher temperatures and is valid for more general cavity geometries, it will be used for the estimation of convection losses in the present receiver design.

Chapter 3

Receiver Prototype

As a proof of concept a CentRec prototype at laboratory scale is designed, built and tested. First experiments are focused on the particle film behavior and its controllability in the "cold" state, where no irradiation is applied. After successfully demonstrating the feasibility of the proposed concept, radiation tests in the High-Flux Solar Simulator (HFSS) [21] at the DLR in Cologne are conducted, paying special attention to the thermal performance of the prototype. Particle outlet temperatures of up to 900 °C are strived for. The experimentally gained data is also used to validate the numerical receiver model, which will be introduced in Chapter 4. A detailed description of the experimental set-up, the used instrumentation and a thorough error analysis are presented in the following.

3.1 Experimental set-up

The test rig as sketched in Figure 3.1 is basically divided into two main parts: the rotating receiver prototype and the stationary infrastructure, which includes the particle feeding and collecting containers, a valve, the receiver drive and the collector. All components are mounted into one frame that consists of easily connectable aluminium profiles. The particle containers are designed for a capacity of 300 kg particles. An electrically driven gate valve, sitting right underneath the feeding container is used to set up various mass flow rates in order to study full and part load conditions.

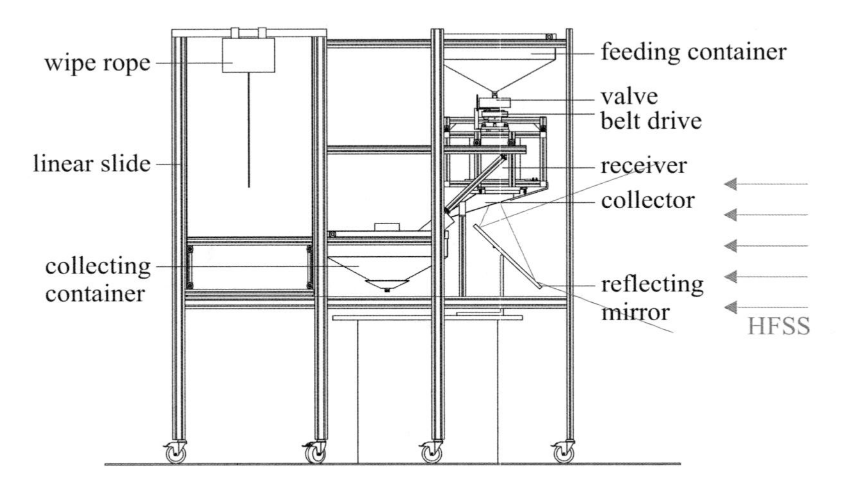

Figure 3.1: Schematic design of the CentRec prototype test rig as tested in the High-Flux Solar Simulator. Set-up is shown for a receiver inclination of $\alpha = 90°$.

Coming from the feeding container the particles enter the receiver where they are heated up by incoming radiation. Collected by the stationary mounted collector, the hot particles are led into a measurement device that continuously measures the particle mass flow rate. For the sake of clarity this instrument is not depicted in Figure 3.1 but a detailed description is given in Chapter 3.2.2. The target flux density has been defined to be $1\,\mathrm{MW\,m^{-2}}$ in the aperture plane. As the maximum input power of the HFSS has been originally estimated to be about 15 kW a circular aperture of 137 mm diameter is calculated and directly integrated within the collector. After leaving the mass flow rate measurement device the particles enter the collecting container where they are cooled down again through convection to the ambience. Since the experiments are run in batch mode a linear slide and two wire ropes are installed for the simple exchange of feeding and collecting tank.

The receiver is mounted in such a way that every inclination angle from 0° to 90° can be studied. As the radiation axis of the HFSS is horizontal, a mirror in 45° position between aperture normal and horizontal for experiments in a 90° configuration is needed to reflect the incoming irradiation into the receiver. Experiments at $\alpha = 45°$ can be conducted without the mirror.

The input power for the high flux tests of the CentRec prototype is provided by the High-Flux Solar Simulator which is based on ten elliptical reflectors with Xenon short-arc lamps. A photo of the HFSS is presented in Figure 3.2. A target area of about $100\,\mathrm{cm}^2$ placed at the lamps' focal length of 3 m can be applied with a maximum radiant power of 20 kW. Flux densities of over $4\,\mathrm{MW\,m^{-2}}$ are possible [21].

Figure 3.2: High-Flux Solar Simulator.

3.1.1 Receiver

A detailed assembly drawing of the prototype is shown in Figure 3.3. The actual heating process of the particles takes place inside a cylindrical cavity of 170 mm internal diameter and 260 mm length. It is concentrically fixed by two holding rings inside the external housing that consists of a stainless steel tube (1.4302). As temperatures of up to 1000 °C are expected, the cavity is made of the Nickel-based alloy Inconel 617 (2.4663) which still exhibits a high tensile strength at higher temperatures [5].

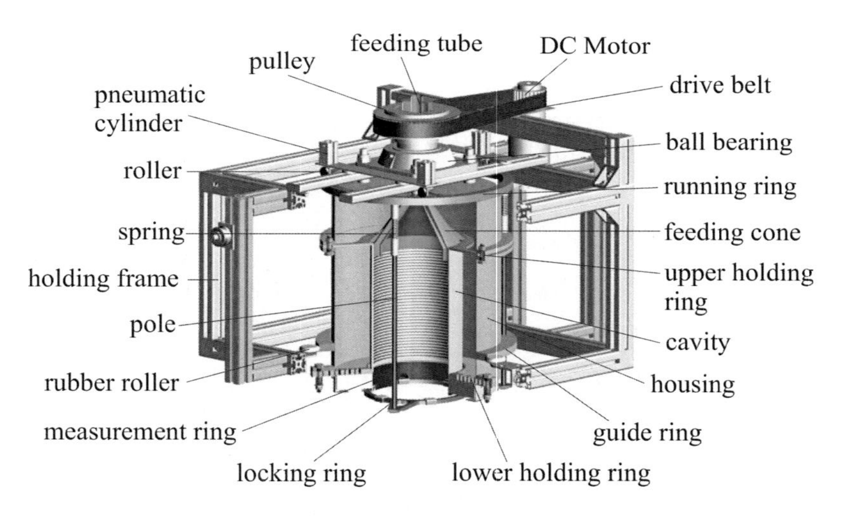

Figure 3.3: Assembly drawing of the CentRec prototype.

Due to the original idea of testing different cavity materials a concept with holding rings is developed to ensure the exchangeability of the cavity. The same Nickel-based alloy is used for the lower holding ring as it can also be heated up to 1000 °C. Since its arms are bolted together with the outer tube, they are exposed to the ambience. Potentially high thermal gradients are the consequences together with high material stresses. To avoid plastic deformations of the lower holding ring its arms are therefore permeated with meanders enhancing the thermal resistance to the outside and homogenizing the temperature distribution.

The upper holding ring is made of 5 mm thick 1.4302 steel as the temperatures at the receiver inlet are expected to be rather low. It is fixed to the flange of the housing and functions as the upper centering of the cavity and the feeding cone. In order to directly determine the particle outlet temperature, an additional ring component is installed right underneath the lower holding ring. Its design and functionality are described in Chapter 3.2.1.

As a part of the actuation system a ball bearing and a pulley are mounted on the outer feeding tube providing the radial and axial bearing of the receiver. The required rotation power is provided by a 200 W DC motor, which is connected to the pulley via a drive belt. Additional radial centering is ensured at the lower third of the receiver. Three hard rubber rollers are fixed with the stationary holding frame and run on a guide ring welded to the external housing.

3.1.2 Particle feeding

When particles enter the cavity it must be ensured that they do not fall through the receiver but stick and slowly move downwards along the wall. Therefore, a double-walled feeding cone (Figure 3.4) is developed to accelerate the particles to the desired circumferential speed before entering the cavity. In the present configuration 16 fins are mounted between the inner and outer cone dividing it into 16 chambers where the particles are accelerated. Due to its rather complex geometry and the necessity of low production tolerances regarding concentricity the feeding cone is manufactured through laser-sintering. The utilized material is a St-70 construction steel.

Particle loss through the open aperture is a critical issue. Especially at the start-up a considerable amount of particle loss can

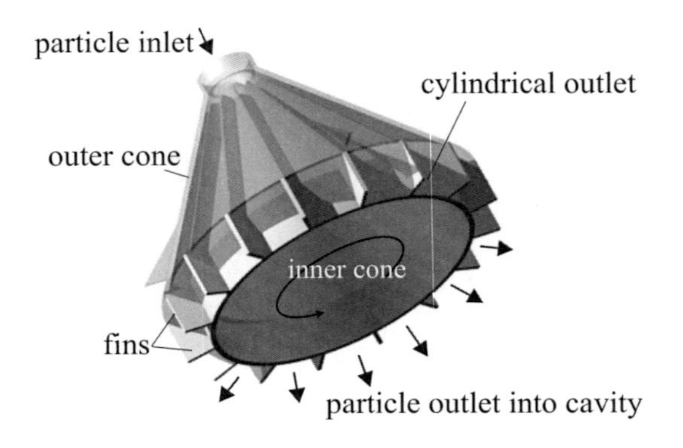

Figure 3.4: Sketch of the feeding cone installed on top of the cavity.

be noticed. A cylindrical part also equipped with fins, that extends into the cavity, completes thus the feeding cone preventing particles to fall through the aperture. Measurements have proven that with the cylindrical outlet the particle loss could be indeed minized to about 0.1 %. For experiments with the prototype this loss is sufficiently low. However, not to affect thermodynamic measurements, for scale-up receivers with bigger diameters and higher mass flow rates the particle losses would need to be much lower.

3.1.3 Insulation

To minimize outward conduction losses microporous insulation based on pyrogenic silica [2] is used to fill the gap between cavity and housing. The required insulation thickness is determined considering the maximal allowed surface temperature on the external housing wall of 60 °C due to the measurement system installed there. The extrapolation of data gained from measurements in Chapter 2.3.4 yields 400 W conduction losses at a cavity wall temperature of 1000 °C. Considering the experimentally determined thermal conductivity of $\lambda_{ins} = 0.175\,\mathrm{W\,K^{-1}\,m^{-1}}$ for the convection test rig [76] and the equation of thermal conduction through cylindrical shells, an insulation thickness of 50 mm is calculated and implemented.

3.2 Instrumentation

For model validation and thermal performance evaluation, quantitative measurements are of crucial importance. The receiver efficiency η_{th} is defined as

$$\eta_{th} = \frac{\dot{Q}_{abs}}{\dot{Q}_{in}}, \tag{3.1}$$

whereas the absorbed heat flow rate by particles $\dot{Q}_{abs}$ is determined by

$$\dot{Q}_{abs} = \dot{m} \int_{T_{in}}^{T_o} c_p(T) dT. \tag{3.2}$$

The basic quantities to be known are therefore the particle mass flow rate $\dot{m}$, its inlet (T_{in}) and outlet temperature T_o, the heat capacity c_p and the incoming heat flow rate $\dot{Q}_{in}$. The following chapters will give a detailed overview of the implemented measurement methods.

3.2.1 Temperature measurement

The temperature distribution along the receiver wall is measured by 3 x 5 plus 3 x 4 thermocouples, equally distributed in six rows, as sketched in Figure 3.5. In order to avoid disturbtion of the particle film on the inner cavity wall and as the cavity is considered to be exchangeable, TCs are installed on the inner wall of the insulation, right behind the cavity. Three additional sensors are installed inside the feeding cone and six more are used to measure the particle outlet temperature. Overall, a total of 36 thermocouples are connected to four TC input modules, mounted on the external surface of the housing. Since the sensors are integrated in the rotating receiver system, a co-rotating wireless LAN router is used to transmit the sampled data to the measuring computer via telemetry. A lithium-polymer battery provides the required power for modules and router. The sampled data is recorded and processed by a LabView routine.

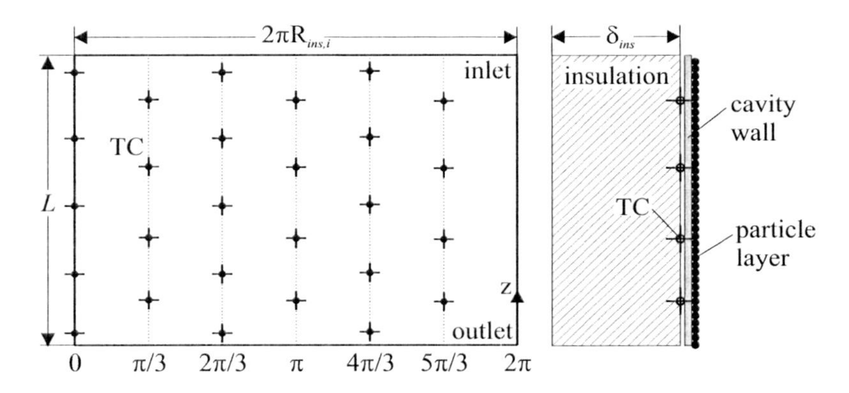

Figure 3.5: Thermocouple positions in the inner insulation wall.

Supplementary to the thermocouple measurements an infrared (IR) camera is used to observe the temperature distribution inside the cavity. IR cameras detect thermal radiation, that is emitted by every body with a temperature $> 0\,\mathrm{K}$. Considering the Stefan-Boltzmann law and the body emissivity the radiated energy can be then converted to a corresponding temperature [49]. However, IR measurements should be taken with caution as the determined temperature is strongly dependent on the body emissivity and the overall radiation situation. In the present experiments neither the particle's emissivity ε is exactly known nor its temperature dependence, which makes a determination of the particle temperature using IR quite inaccurate.

Moreover, reflected radiation inside the cavity is also detected that could additionally falsify the true temperature. Recordings with the IR camera are nevertheless taken to gain insight of the temperature distribution of the particle film and to compare them qualitatively to measurements with thermocouples.

For the determination of the receiver efficiency it is essential to precisely measure the particle outlet temperature. Due to space limitations it was not possible to insulate the collector or the container, where the particle temperature could have been easily measured. An alternative concept had to be developed instead, in order to measure the temperature of the particles directly at the receiver outlet.

For the correct contact-based measurement of bulk material temperature the radius of the bulk volume should be at least five times bigger than the sensor diameter [42]. Extensive studies were undertaken [62] resulting in a measurement system, that co-rotates with the receiver and measures the particle temperature in defined time periods.

The so-called Temperature Measurement Ring (TMR), sketched in Figure 3.6, basically consists of a sheet ring with six radially outwards pointing chambers. Thermocouples are placed in the middle of each chamber, where particles are enclosed. The chambers' dimension is such that the critical particle volume for an accurate temperature measurement is ensured. However, once the chambers are filled up with particles, they will cool down in a short time. To be still able to measure the particle outlet temperature over a certain time range cold particles need to be released and the chambers refilled with new hot particles.

This is realized with the locking ring sitting right underneath the chambers. In the closed position, the chambers are filled with particles and their temperature is recorded. Once the chambers are full, the locking ring is pulled down by three pneumatic cylinders, as seen in Figure 3.3, and cooled down particles are released.

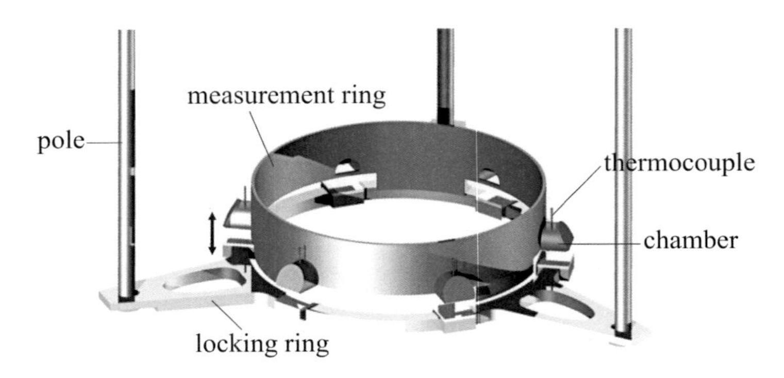

Figure 3.6: Schematic of the TMR, developed in order to measure the particle temperature directly at the receiver outlet [62].

The locking ring is connected via three poles and a running ring to the stationary mounted pneumatic cylinders. Rollers installed on the cylinder axes and running on the ring serve as the connection between rotational and stationary system. After being pulled down, the locking ring is pushed back to its initial position by three return springs and the chambers are closed again. The pneumatic cylinders are automatically triggered via a control program in LabView, where the time between each emptying period can be predefined. For all subsequent measurements, chambers are opened every three to five minutes for one second, depending on the corresponding mass flow rate.

An exemplary time distribution of the measured temperature in one chamber is plotted in Figure 3.7. The periodic opening and closing of the chamber is clearly revealed. Particles are released at the low point of each cycle. After being opened for one second, the empty chamber is filled with new hot particles and the temperature rises until the chamber is full. The actual particle temperature is therefore expected to be at the curve maximum. The decrease in temperature indicates the cooling process of the particles in the filled chamber.

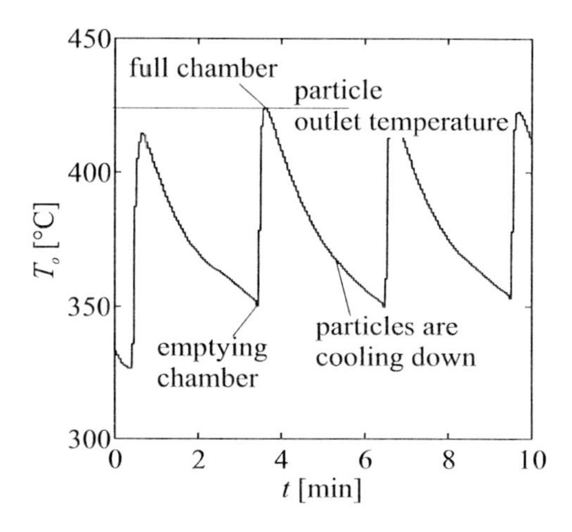

Figure 3.7: Typical distribution of particle outlet temperature measured by the TMR. The measured behavior of one chamber is plotted.

3.2.2 Mass flow rate measurement

To investigate the receiver performance at various load conditions, the particle mass flow rate is altered by placing different masks with predefined orifice diameters D_{or} between valve and feeding tube. The orifice sizes are calibrated in advance, whereas each D_{or} corresponds to one specific mass flow rate. Diameters from 5 to 10 mm are correlated to mass flow rates from 2 to $16\,\mathrm{g\,s^{-1}}$.

An additional measurement system between collector and collecting container is installed to monitor on-line the mass flow rate during each experiment. A schematic is presented in Figure 3.8. As very low mass flow rates of up to $2\,\mathrm{g\,s^{-1}}$ are operated scales with a resolution of at least 1 g are required. Affordable scales in this range are limited to a maximum weight, why an intermediate container with a loading capacity of about 8 kg is integrated.

To avoid overflow, it can be opened and closed by a valve that is automatically controlled by a pneumatic cylinder. When the reservoir is full, the valve is opened and the particles are released into the collecting container. After the intermediate container is emptied, the valve is closed and it can be filled with new particles again. Similar to the mechanism of the TMR the pneumatic cylinder is controlled by a measurement program where the open to close cycle can be set individually according to the mass flow rate. The filling process is monitored and the weight is transmitted to the computer every six seconds. The derivative of the mass-to-time curve determines the mass flow rate at every timestep. Any fluctuations of the particle movement as they pass through the cavity can thus be observed directly.

The intermediate container is also made of high temperature Nickel-based alloy as it collects the heated particles. Sufficiently thick insulation is implemented protecting the pneumatic cylinder and scales from heat.

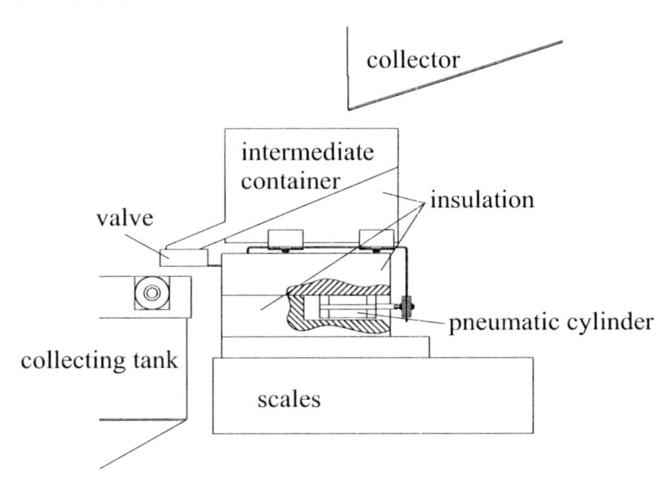

Figure 3.8: Schematic of the mass flow rate measurement system placed between collector and collecting tank.

3.2.3 Input power measurement

Incoming radiation is measured using the FATMES system, which was developed for flux measurements in the solar furnace in Cologne [51]. It consists of an air-cooled target plate coated with a white and diffusely reflective coating (Al_2O_3). A video camera looking through a set of ND filters records the image of the concentrated beam as it is reflected on the target. A camera calibration factor is calculated by comparing the absolute flux density measured with a point measurement flux gauge at the specific point with the camera grey scale intensity.

Due to space limitations it was not possible to install a moving target and measure the incoming heat flux before each single experiment. Instead, the flux distribution in the receiver aperture plane is recorded for all relevant lamp configurations at one receiver inclination before the actual experiments. The overall radiation power entering the receiver is calculated by integrating the flux densities over the aperture area.

3.2.4 Rotation rate measurement

The receiver rotation speed is accurately measured via a rotary encoder directly mounted on the rotating axis of the DC motor. A frequency counter converts the encoder impulses into voltage signals which are transferred to the data aquisition system. The resolution of the encoder is defined with 500 impulses per revolution taking into account the overall gear ratio between motor and receiver of 40:3.

The DC motor is powered by a thyristor drive whose input voltage signal is set by the measurement program. As the rotation speed increases for a constant voltage signal due to alteration of the inner motor resistance and assumedly decreased friction, a PID controller is implemented in the measurement program in order to ensure constant receiver rotation.

According to manufacturer's information, the relative error of the overall system is assumed to be below 1 % for considered rotation speeds between 150 and 190 rpm. Comparative measurements to an optical measurement method for example, have even revealed maximum errors of only 0.5 %.

3.3 Error analysis

Experimentally determined quantities always exhibit some uncertainties. The result of any measurement of a quantity q is stated as

$$(\text{measured value of } q) = q_{best} \pm \delta q, \tag{3.3}$$

with q_{best} as the best estimate of q and δq as the absolute uncertainty in the measurement [72].

When the quantity of interest cannot be directly measured but derived from the calculation of quantities which can be directly measured as in the case of (3.1), the resulting uncertainty is found by the general formula for error propagation

$$\delta q = \sqrt{\left(\frac{\partial q}{\partial x}\delta x\right)^2 + ... + \left(\frac{\partial q}{\partial z}\delta z\right)^2}. \tag{3.4}$$

The equation is only valid for small, independent and random uncertainties.

Uncertainties are in general categorized in two groups: the random uncertainties δq_{ran}, which can be treated statistically and reduced by increasing the number of measurements, and the systematic ones δq_{sys}, which cannot be diminished [72]. Random uncertainties are defined as the standard deviation of the mean (SDOM)

$$\delta q_{ran} = \frac{1}{\sqrt{N}}\sqrt{\frac{1}{N-1}\sum(q_i - \overline{q})^2}, \tag{3.5}$$

where the best estimate is denoted as the mean of N measurements

$$\overline{q} = \frac{\sum q_i}{N}. \tag{3.6}$$

Systematic errors however are hard to identify as they are not revealed by repeating the measurements, but mainly based on factors like the imperfection of the measurement devices, their influence on the experiment or surrounding effects. If an estimation of the systematic uncertainty can be made, a reasonable (but not rigorously justified) expression of the total uncertainty is

$$\delta q_{tot} = \sqrt{(\delta q_{ran})^2 + (\delta q_{sys})^2}. \tag{3.7}$$

3.3.1 Temperature uncertainty

For the determination of the receiver efficiency there are two essential temperatures: The particle inlet and outlet temperature, T_{in} and T_o. Both temperatures are measured by 1 mm diameter thermocouples of type N, which exhibit according to the DIN standard DIN EN 60584 an uncertainty of ± 2.5 °C or 0.75 % of the measured value. The used data logger transmitting the temperature to the processing computer is specified with a precision of 0.1 % by the manufacturer. A calibration of the thermocouples with a calibration kiln [1] have revealed a maximum overall error of 0.68 %, which is used as the systematic uncertainty δT_{TC} of the installed thermocouples.

As the thermocouple measuring T_{in} is directly mounted inside the feeding container, where the probe tip is completely surrounded by a sufficient amount of particles, the total uncertainty of the inlet temperature is only composed of δT_{TC} of the measurement devices. Statistical errors are neglected as the sampling rate of 1 Hz is sufficient high for a minimum measuring period t_p of 20 minutes.

The particle outlet temperature is measured with the TMR as described in Chapter 3.2.1. To get the overall outlet temperature T_o of each experiment, the mean value of the six chambers $\overline{T}_{ch}$ and the averaged value of peak temperatures $\overline{T}_{peak}$ during t_p must be taken into account as

$$T_o = \overline{T}_o = \frac{1}{N_{ch}} \sum \left(\frac{\sum T_{peak,i,j}}{N_{peak}} \right)_{ch,j} . \tag{3.8}$$

With the SDOM of $\overline{T}_{ch}$ and $\overline{T}_{peak}$ according to (3.5), the total random uncertainty of T_o is calculated as

$$\delta T_{o,ran} = \sqrt{(\delta \overline{T}_{ch})^2 + (\delta \overline{T}_{peak})^2}. \tag{3.9}$$

The outlet temperature of the particles is determined by the peak value in the temperature measurement with the TMR as described in Chapter 3.2.1. The time at the peak value can be also considered as the time where the chamber is entirely filled with particles after an opening-closing period of the TMR. As the filling process of the chamber takes about 10 to 20 seconds, depending on the corresponding mass flow rate, particles, which are already inside the chamber, are cooled down again. A mixed temperature in the chamber is thus reached leading to the assumption that the measured peak value might be lower as the actual particle outlet temperature. In order to roughly estimate this systematical error $\delta T_{ch,sys}$ the filling time of the chamber is determined by measuring the time between opening the TMR and the measured peak value. Applying this time to the subsequent cooling curve of the particles inside the chamber, the temperature difference that occurs in this time intervall is evaluated to be the systematical error.

The final uncertainty δT_o is determined by (3.7) with $\delta T_{o,ran}$, δT_{TC} and $\delta T_{ch,sys}$. Additional systematic uncertainties due to imprecise TC positioning are omitted as it is assumed that the thermocouple is fully covered by particles when the chamber is filled completely.

For validation purposes of the numerical model the temperature distribution along the receiver wall is considered. To avoid the interference of the particle movement, thermocouples are not mounted directly on the inner wall surface, but placed behind the cavity on the inner insulation wall (described in Chapter 3.2.1). Additional systematic errors must be thus taken into account. In order to estimate these uncertainties, experiments are conducted where heating wires are used to heat up the cavity. Its inner wall surface is equipped with extra thermocouples measuring the actual wall temperature as a reference. While no rotation and no particles are applied, the temperature determined by the additional thermocouples on the inner cavity wall surface are compared to the measurements by the thermocouples installed on the inner insulation wall. The maximum discrepancy between both measurements is designated as the systematic uncertainty of the temperature distribution along the receiver wall, which is about 10 K for considered temperature levels from 200 to 500 °C.

3.3.2 Mass flow rate uncertainty

The uncertainty of the measured mass flow rate is solely considered for the measurement system installed at the receiver outlet. It consists mainly of statistical errors during the measurement periods and the systematic error of the weighing scale, which is about 1 g according to manufacturer's data. As $\dot{m}$ is calculated by $\dot{m} = \Delta m / \Delta t$ error propagation must be considered leading to

$$\delta \dot{m}_{sys} = \sqrt{\left(\frac{1}{\Delta t} \delta m_{sys}\right)^2 + \left(\frac{m}{(\Delta t)^2} \delta(\Delta t)\right)^2} \qquad (3.10)$$

as the systematic uncertainty. The error of Δt for each measurement of m is neglected as it is set by the measurement program.

Consider statistical uncertainties evaluated for each experiment according to (3.5), the overall mass flow rate error is also calculated using (3.7).

3.3.3 Uncertainty of the input power

The incoming heat flux distribution into the receiver is determined as described in Chapter 3.2.3. According to the HFSS operator the measurement uncertainty of the input power is about 3 to 4 % [77]. However, since the heat flux measurement is decoupled from the actual experiments additional errors occur due to positioning and set-up uncertainties resulting from the measurement procedure. In a first step, the receiver is mounted to its desired position and the aperture position is marked by two laser-pointers. Secondly, the receiver is removed and the target plane is installed at the marked position. The lamps of the HFSS are focused on the target and the resulting flux densities are measured. The target is removed again and the receiver is positioned back. Despite being highly careful during the mounting processes, positioning errors are still likely to occur. As the calculation of the input power is done by integrating over the aperture area, slight deviations of just one to two centimetres of its position in the heat flux maps lead to calculation differences of several percent. Accounting for these possible positioning uncertainties an overall error of $\delta \dot{q}_{in} = 5\,\%$ is thus estimated.

3.3.4 Overall uncertainty

The total heat flow rate $\dot{Q}_{abs}$ absorbed by the particles is calculated according to (3.2). Here, the temperature dependent heat capacity c_p of the particles is taken from the measurements by Siegel [65], given in Appendix A. Comparison to literature data for corundum [11, 73] reveals good agreement. In order to be able to calculate $\dot{Q}_{abs}$ analytically, the values of c_p are fitted by a polynomial of fourth order, for which the integral is quite easy to solve. The fitted heat capacity is denoted as

$$\begin{aligned} c_p(T) = [&- 2.853 \times 10^{-9}(T/{}^\circ\mathrm{C})^4 + 7.059 \times 10^{-6}(T/{}^\circ\mathrm{C})^3 \\ &- 5.795 \times 10^{-3}(T/{}^\circ\mathrm{C})^2 + 2.439(T/{}^\circ\mathrm{C}) \\ &+ 677]\ \mathrm{J\,kg^{-1}\,{}^\circ C^{-1}}, 25\,{}^\circ\mathrm{C} \leq T \leq 1000\,{}^\circ\mathrm{C}. \end{aligned} \tag{3.11}$$

For the uncertainty analysis of $\dot{Q}_{abs}$, the error in the c_p measurement has to be considered as well, which is denoted by around 2 % according to manufacturer's data for the measurement device [3]. To simplify the error calculation a mean value of c_p is hence determined with

$$\bar{c}_p = \frac{\int_{T_{in}}^{T_o} c_p(T)dT}{T_o - T_{in}}. \tag{3.12}$$

While $\dot{Q}_{abs}$ is evaluated by (3.2), its uncertainty is derived according to (3.4), such that

$$\delta\dot{Q} = \sqrt{\left(\frac{\partial\dot{Q}}{\partial\dot{m}}\delta\dot{m}\right)^2 + \left(\frac{\partial\dot{Q}}{\partial c_p}\delta c_p\right)^2 + \left(\frac{\partial\dot{Q}}{\partial T_{in}}\delta T_{in}\right)^2 + \left(\frac{\partial\dot{Q}}{\partial T_o}\delta T_o\right)^2}.$$

Due to clarity reasons, $\dot{Q}$ is used as an abbreviation of $\dot{Q}_{abs}$. Considering (3.1) for the calculation of η_{th}, the absolute uncertainty of the experimentally determined thermal receiver efficiency is finally calculated as

$$\delta\eta_{th} = \sqrt{\left(\frac{1}{\dot{Q}_{in}}\delta\dot{Q}_{abs}\right)^2 + \left(\frac{\dot{Q}_{abs}}{\dot{Q}_{in}^2}\delta\dot{Q}_{in}\right)^2}. \tag{3.13}$$

Individual errors of each experiment are given in Appendix B.

Chapter 4

Numerical Receiver Model

For the overall efficiency analysis of a CSP system knowing the thermal receiver efficiency is of crucial importance. In order to characterize the thermal performance of the CentRec concept a three-dimensional, steady-state numerical model of the laboratory prototype is developed. It should depict the governing physical mechanism in a sufficient level of detail, while keeping the computational effort to a reasonable amount. The basic thermal radiation, conduction, convection and optical losses are calculated.

The present model is built in the commercial software ANSYS using the finite element method (FEM). In the following sections the model approach, its basic structure and applied boundary conditions are introduced and first validation cases are discussed.

4.1 Model simplifications

The model geometry is a simplified version of the CentRec prototype. The main receiver part as described in Chapter 3.1.1, consisting of cavity, insulation and external housing, is modeled as one cylindrical tube. Its inner diameter equals the cavity diameter $D_{cav} = 170\,\text{mm}$ and its outer diameter equals the diameter of the housing (outer wall) $D_{ow} = 315\,\text{mm}$. The particle inlet part, including insulation, the upper external housing and the double-walled feeding cone, is also simplified to a cylindrical block with a conical recess in the dimensions of the feeding cone inside. The TMR is included in the model as an additional cylindrical tube mounted directly below the main receiver body. As its effect on the thermal receiver simulation is considered to be of minor importance the modelling of the full collecting ring geometry is omitted. It is replaced by a 2D circular ring (front ring) with its inner diameter corresponding to the aperture diameter $D_{ap} = 138\,\text{mm}$ and its outer diameter to D_{cav}. For energy balance purposes the ambience is modelled in the aperture plane as a 2D circular surface which closes the cavity model. A half-section of the numerical model is sketched in Figure 4.1.

For the model development the ideal case is assumed, where the receiver wall is fully covered by a single-layered, optically dense and homogeneously moving particle film. As the interaction of single particles is not of any interest for the present simulation, the modelling of discrete particles is neglected. Instead, the entire particle film is represented by continuous flowing, so-called "fluid lines", where heat is transferred due to mass transport of the fluid. Directly connected to the cavity walls, the fluid lines are gradually heated up by incoming radiation. The resulting end temperature of the fluid equals the expected particle outlet temperature. The flow velocity is defined by the corresponding experimental particle mass flow rate and the fluid properties are set identically to the particle properties. An exemplary fluid line is denoted by the orange line sketched in Figure 4.1.

In order to consider the receiver rotation without actually rotating the model, the fluid lines are each arranged on a helical

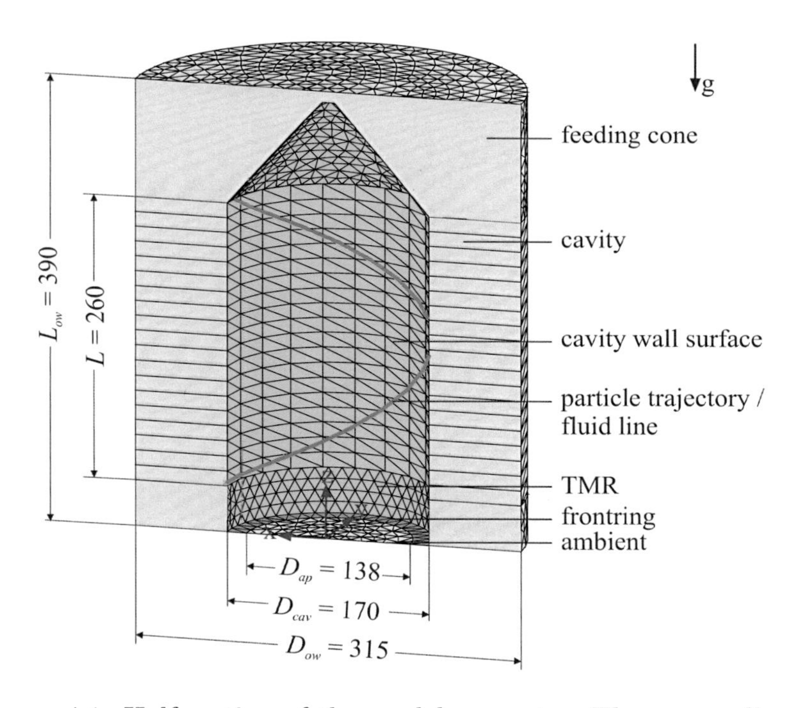

Figure 4.1: Half-section of the model geometry. The orange line denotes the particle trajectory, which is modelled as a so-called "fluid line". Dimensions are in mm.

path which simulates the particle trajectory in the receiver from a stationary point of view. In the numerical model, the entire cavity wall is covered by adjacently placed helical fluid lines modelling the heat transport of the particle film. As the simulation is focused on the direct heating of the particles on the cavity wall, their acceleration and potential preheating by the feeding cone are neglected. Particles are solely modelled on the cavity wall.

4.2 Governing equations

To precisely calculate the receiver efficiency η_{th} the accurate modelling of the three basic thermal losses consisting of conduction ($\dot{Q}_{cond}$), convection ($\dot{Q}_{conv}$), radiation ($\dot{Q}_{rad}$) and optical losses ($\dot{Q}_{opt}$) is of crucial importance, since

$$\eta_{th} = 1 - \frac{(\dot{Q}_{rad} + \dot{Q}_{cond} + \dot{Q}_{conv}) + \dot{Q}_{opt}}{\dot{Q}_{in}}. \tag{4.1}$$

4.2.1 Conduction and convection

Derived from the first law of thermodynamics, the conservation of thermal energy, the general heat equation is defined as [20]

$$\rho c \left(\frac{\partial T}{\partial t} + \boldsymbol{v} \mathrm{grad} T \right) = -\mathrm{div} \dot{\boldsymbol{q}} + \dot{S} \tag{4.2}$$

in the conservative form with ρ as the body density, $c = c(T)$ as the temperature dependent specific heat capacity, $\boldsymbol{v}$ as the velocity vector for mass transport of heat and $\dot{S}$ as the internal heat generation rate per unit volume. Consider a three-dimensional, steady-state case with no internal heat sources and Fourier's law

$$\dot{\boldsymbol{q}} = -\lambda \mathrm{grad} T, \tag{4.3}$$

(4.2) is reduced to the differential equation

$$\rho c \left(u\frac{\partial T}{\partial x} + v\frac{\partial T}{\partial y} + w\frac{\partial T}{\partial z} \right) + \frac{\partial}{\partial x}\left(\lambda \frac{\partial T}{\partial x} \right) + \frac{\partial}{\partial y}\left(\lambda \frac{\partial T}{\partial y} \right) + \frac{\partial}{\partial z}\left(\lambda \frac{\partial T}{\partial z} \right) = 0 \tag{4.4}$$

written in the non-conservative form. $\lambda = \lambda(T)$ denotes the temperature dependent thermal conductivity for an isotropic material. The mass transport of heat through the fluid lines is represented by the first term on the left.

To minimize the computational effort, no air flow is modeled as it is done in Computational Fluid Dynamic (CFD) simulations for the calculation of convective heat losses. Surface convection is applied instead as a boundary condition within solving the heat equation.

4.2.2 Radiation

Because of the rough particle film surface and its fairly wave length independent optical properties (see Appendix A) grey diffuse radiation exchange can be assumed. The radiation exchange for an enclosure of N gray diffuse surfaces is described by the equation system [67]

$$\sum_{i=1}^{N} \left(\frac{\delta_{ji}}{\varepsilon_i} - F_{ji}\frac{1-\varepsilon_i}{\varepsilon_i} \right) \frac{\dot{Q}_i}{A_i} = \sum_{i=1}^{N} (\delta_{ij} - F_{ji})\sigma T_i^4 . \tag{4.5}$$

Corresponding to each surface, i takes one of the values 1,2,...,N, and δ_{ji} is the Kronecker delta, defined as

$$\delta_{ji} = \begin{cases} 1 & \text{when } j = i \\ 0 & \text{when } j \neq i. \end{cases}$$

The effective emissivity is denoted by ε, the surface area by A and the radiative heat loss by $\dot{Q}$. The radiation view factor F_{ji} is defined as

$$F_{ji} = \frac{1}{\pi A_j} \int_{A_j} \int_{A_i} \frac{\cos\beta_i \cos\beta_j}{r^2} dA_j dA_i \tag{4.6}$$

with r as the distance between the radiative surfaces and β as the angle between the surface normal and the distance line.

4.3 Numerical method

As no analytical solution of (4.4) exists, the solution is approximated using the finite element method, where the computational domain is discretized in a finite number of smaller subdomains. The solution of simpler equations of these elements by known solution techniques are united to approximate the final solution. Advantages of FEM are, amongst others, the calculation of complex and discontinuous geometries and structures, the capture of local effects and a modular composition. In a thermal FEM analysis, the considered degree of freedom (DOF) is the temperature.

4.3.1 Boundary conditions

A schematic of the applied boundary conditions and heat transfer mechanisms is presented in Figure 4.2. Equation (4.4) is supplemented by Dirichlet boundary conditions for the temperature at the ambient interface, such that $T|_{\infty} = 25\,°\mathrm{C}$.

At the entire inner wall surface of the receiver, including feeding cone and TMR, a von Neumann boundary condition is implemented for the temperature with

$$-\lambda \mathrm{grad} T \cdot \boldsymbol{n}|_{wall} = \dot{q}_{in}. \tag{4.7}$$

The normal surface vector is denoted by $\boldsymbol{n}$. The calculation of the incoming heat flux $\dot{q}_{in}$ from the solar simulator via ray-tracing is explained in detail in Chapter 4.5.

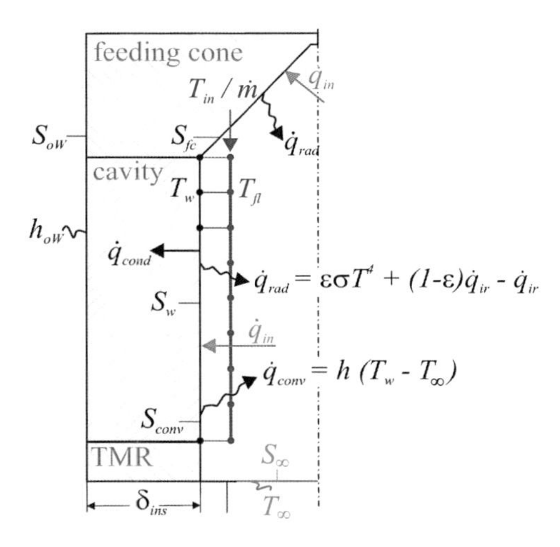

Figure 4.2: Schematic of applied boundary conditions and heat transfer mechanisms. For clarity reasons the helical flow pipe is sketched as a straight line (blue).

The outer receiver wall is specified with a convective boundary condition, such that

$$-\lambda \mathrm{grad} T \cdot \boldsymbol{n}|_{ow} = h_{ow}(T_{ow} - T_\infty). \tag{4.8}$$

The heat transfer coefficient is chosen to be $h_{ow} = 10\,\mathrm{W\,m^{-2}\,K^{-1}}$ according to a value for free convection on a vertical cylinder [7].

For the calculation of convection losses, convective boundary conditions are applied on the predetermined surface of the convective area S_{conv},

$$-\lambda \mathrm{grad} T \cdot \boldsymbol{n}|_{conv} = h(T_w - T_\infty). \tag{4.9}$$

The heat transfer coefficient h is derived from Clausing's correlation [18] according to the findings in Chapter 2. The convective surface area is calculated depending on the receiver inclination angle. Clausing's formula is however only valid for $\alpha < 90°$ as he assumed no convection losses at $\alpha = 90°$. Several subsequent experiments [71, 76] have proved his statement to be inaccurate wherefore h for the face-down configuration is estimated for an inclination of $\alpha = 89.9°$ instead.

As h is calculated depending on the surface temperature of the convective area, which is not known at the beginning, the correct heat transfer coefficient must be found iteratively. With an initial guess of the expected mean surface temperature $\overline{T}_{conv,0}$ a random h_0 is defined. After a first complete run with h_0, a new temperature $\overline{T}_{conv,1}$ can be evaluated followed by a calculation of h_1. These iterations are repeated until a heat transfer coefficient h_n is found, that satisfies the convergence criteria of $\left|\overline{T}_{conv,n} - \overline{T}_{conv,n-1}\right| \leq 1\,°\mathrm{C}$. Preliminary test runs have revealed that in general one iteration is sufficient to achieve the required convergence criteria.

Consider the fluid elements denoted by the blue line in Figure 4.2, the particle inlet temperature T_{in} is employed as an initial condition on the first node of each discretized fluid line. Each line element is applied with a corresponding mass flow rate $\dot{m}_{fl}$, which is the full particle mass flow rate $\dot{m}$ divided by the number of representative overall fluid lines. The values of $\dot{m}$ and T_{in} are defined according to the experiments and vary depending on the experimental conditions.

4.3.2 Material properties

Defining an accurate thermal conductivity coefficient is of crucial importance for the solution of the heat equation. Although the conductivity of the used insulation material λ_{ins} is known, it does not represent the entire conductivity of the prototype,

λ_{prot}, as heat losses due to gaps and thermal bridges through the holding rings need to be considered. Experiments similar to those described in Chapter 2.3.4 are thus conducted in order to determine an overall λ_{prot}. By combining simulations and experimental data, a temperature independent conductivity value has been evaluated and implemented with $\lambda = 0.168\,\mathrm{W\,m^{-1}\,K^{-1}}$. A detailed description of the semi-empirical procedure can be found in Appendix A.

For the accurate simulation of optical and radiation losses, two additional material properties of the particles need to be considered: The absorptivity a and the emissivity ε. Detailed measurements of a and ε of the considered particles, conducted by Siegel [65], are presented in Appendix A. Despite the visible dependence of a and ε on the wavelength, but out of convenience, subsequent simulations are performed with constant, wavelength independent values with $a = 0.89$ and $\varepsilon = 0.82$. To estimate the induced error by assuming constant values, a thorough sensitivity analysis regarding the influence of a and ε on the simulation results is given in Chapter 6.3.1.

In order to evaluate the heat absorbed by the particles, the temperature dependent heat capacity of the particles is applied corresponding to the values already introduced in Chapter 3.3.4.

The remaining parts of the receiver model, such as feeding cone, TMR and front ring, are considered to consist of common steel, with an absorptivity of $a = 0.2$ and emissivity of $\varepsilon = 0.5$. The ambient interface is applied with $a = \varepsilon = 1$, as it is assumed to fully absorb radiation without reflecting anything back.

4.3.3 Finite element approximations

Taking the boundary conditions into account, applying the *Virtual Temperature Principle* and considering a finite element approximation $T = \boldsymbol{N}^T \boldsymbol{T_e}$ with $\boldsymbol{N}$ as an element shape function and $\boldsymbol{T_e}$ as the nodal temperature vector, yield a set of simultaneous linear equations

$$([K_{e,\lambda}] + [K_{e,c}] + [K_{e,mt}]) \cdot \boldsymbol{T_e} = \dot{\boldsymbol{Q}}_{e,q} + \dot{\boldsymbol{Q}}_{e,c} + \dot{\boldsymbol{Q}}_{e,fl} \qquad (4.10)$$

which can be solved. $[K_{e,\lambda}]$ is the element conductivity matrix by diffusion, $[K_{e,h}]$ by surface convection and $[K_{e,mt}]$ by mass transport. The incoming heat flow rate vector is denoted by $\dot{\boldsymbol{Q}}_{e,q}$, the heat flow rate vector by surface convection by $\dot{\boldsymbol{Q}}_{e,c}$ and $\dot{\boldsymbol{Q}}_{e,fl}$ considers the heat transported in the fluid line. A detailed derivation can be seen in the ANSYS Theory Reference [4].

Equation (4.10) is solved using a direct elimination process. The default solver in ANSYS Mechanical is the Sparse Direct Solver, that decomposes the conductivity matrix $[K]$ into lower and upper triangular matrices $[K] = [L][U]$ by assuming sparsely populated finite element matrices [4]. Forward and back substitutions using $[L]$ and $[U]$ are then made to compute the solution vector $\boldsymbol{T_e}$.

The Radiosity Solution Method of ANSYS [4] is used for radiation calculation. Rearrange (4.5) and express it in terms of outgoing radiative fluxes (radiosity) for each surface $\dot{q}_j^0$ and the net flux from each surface $\dot{q}_i$, a set of linear algebraic equations is formed

$$[A]\dot{\boldsymbol{q}}^{\boldsymbol{0}} = \boldsymbol{D} \qquad (4.11)$$

with $A_{ij} = \delta_{ij} - (1 - \varepsilon_i)F_{ij}$ and $D_i = \varepsilon_i \sigma T_i^4$. It is solved by the Newton-Raphson procedure. Here, $[A]$ denotes a full matrix due to the surface to surface coupling represented by the view factors. When the radiosity values are known, the net flux at each surface can be evaluated and be provided as a boundary condition to the finite element model of the conduction calculation. Equation (4.11) is solved iteratively coupled with (4.10) until convergence of the radiosity flux and tenperature is reached for each time step.

View factors are derived by the hemicube method which is based upon Nusselt's hemisphere analogy [19]. It says, that for any surface covering the same area projected on an imaginary hemisphere an identical view factor exists. In the hemicube method an imaginary cube is constructed instead of a sphere. A radiating element face dA_1 is projected onto the five planar surfaces of the hemicube which are disretized into small square elements, called pixels. The view factors from dA_1 to each pixel is precomputed and the factor of the projected patch on the hemicube is found by summing the view factors for the pixels contained within the patch.

4.4 Discretization

In order to model the helical path of the fluid lines, the receiver model is discretized in a specific way which is described in the present chapter. The grid resolution is defined by the number of representative flow pipes, circumferentially distributed, and their number of revolutions.

4.4.1 Particle trajectory

At sufficiently high rotational speed, a single particle in the rotating cavity is forced against the wall by centrifugal acceleration. Due to gravitation and depending on rotation speed and wall roughness the particle slowly moves down. From a stationary, external point of view, the particle describes a helical trajectory from the top to the bottom.

The characteristics of a helix are defined by two parameters: first, the axial distance the helix winds during one full revolution (2π), called the lead. And second, the helix angle which is the angle between any helix and its lead [6].

In the present case, the lead H is derived by dividing the receiver length L by the number of particle revolutions n_{rev},

$$H = \frac{L}{n_{rev}}. \tag{4.12}$$

The according helix angle is calculated as

$$\chi = \arctan\left(\frac{2\pi R}{H}\right). \tag{4.13}$$

Choosing the number of representative fluid lines N_{fl} and therefore the discretization in circumferential direction (φ) determines the width of each cell ($d\varphi$), since

$$d\varphi = \frac{2\pi R}{N_{fl}}. \tag{4.14}$$

Combined with the helix angle, the height dz of one element is defined as

$$dz = \frac{R}{\tan\chi} d\varphi. \tag{4.15}$$

With dz, the number of elements per fluid line is specified and thus the discretization in axial direction (z).

4.4.2 Cavity

The discretization of the cavity mesh in the previous section leads to quadrangular elements. However, in order to accurately model the cylindrical shape of the cavity, prism-shaped elements are more appropriate in ANSYS [4]. Their form and labelling is detailed in Figure 4.3. The base area of each prism is a triangle with exactly defined vertices.

Due to simplicity reasons the mesh generation is done in a polar coordinate system. The principle approach is sketched in Figure 4.4.

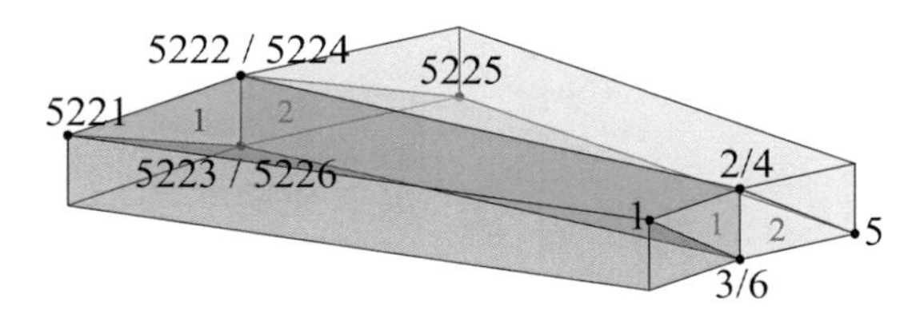

Figure 4.3: Nodes and elements labelling of prism-shaped elements in ANSYS.

The coordinates of each node of the cavity grid are defined as

$$N_{ij,E}(\varphi|z) = N_{ij,E}((i-1)d\varphi|(j-1)dz),$$

with i and j as running indices in φ and z, respectively. E denotes the current element number, N_φ the overall number of nodes in φ and N_z the overall number of nodes in z. The nodes of the outer cavity wall are defined in the same way, just taking the outer wall radius R_o into account. The length L_e of each element thus equals the receiver insulation thickness. Conduction is only considered in longitudinal direction of the element. Preliminary simulations have revealed that discretizing the radial direction of the receiver in just one element is sufficiently accurate for conduction loss calculations through the insulation.

For the heat transfer between cavity wall and fluid lines the wall elements need to be arranged in such a way that they contour the same helical paths as the fluid elements as denoted by the blue area in Figure 4.4. In the grouping routine there are always two adjacent triangular elements per row taken. Their order is

$$E_{ij,1} = 1 + 2(i-1) + 2(j-1)n_{fl}$$

and

$$E_{ij,2} = 2 + 2(i-1) + 2(j-1)n_{fl}.$$

The flow pipes are generated as "line elements" in the middle of the triangular wall elements. The definition of so-called "key-

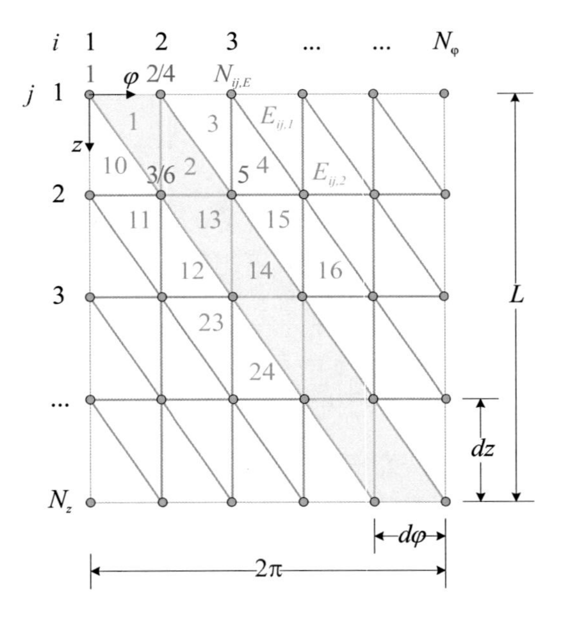

Figure 4.4: Definition of node positions in polar coordinates.

point" positions is sketched in Figure 4.5. The corresponding polar coordinates are determined as

$$K_{ij}(\varphi|z) = K_{ij}(i\frac{d\varphi}{2}|(j-1)dz).$$

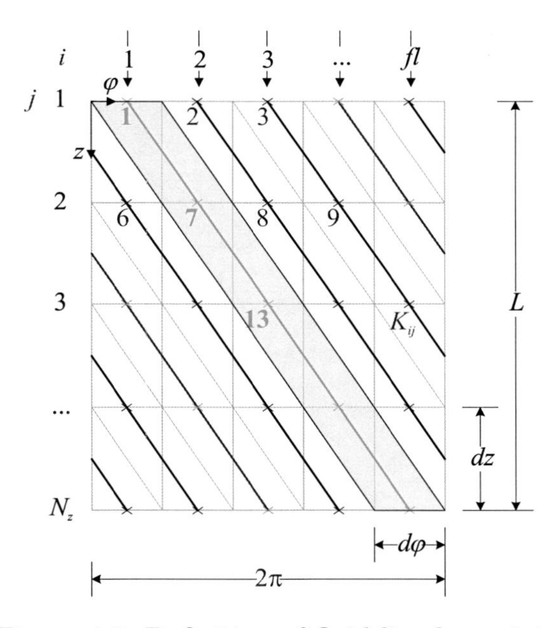

Figure 4.5: Definition of fluid line keypoints.

Similar to the grouping of the wall elements, the keypoint numbers must be assembled and arranged on a helical path as well. They are sorted as

$$K_{ij} = i + (j-1)(n_{fl}+1)$$

per fluid line.

4.5 Incoming heat flux

The implemention of the incoming heat flux distribution is divided into several steps. The basic approach is schemed in Figure 4.6. After the generation of the receiver geometry in ANSYS, a special macro, FEMRAY [75], is used to generate additional elements according to the model geometry and mesh disretization. Surfaces considered in the raytracing calculation are selected and those with the same optical properties are grouped together. According to the given absorption coefficients for each component (Chapter 4.3.2) optical losses are directly evaluated in the heat flux calculation.

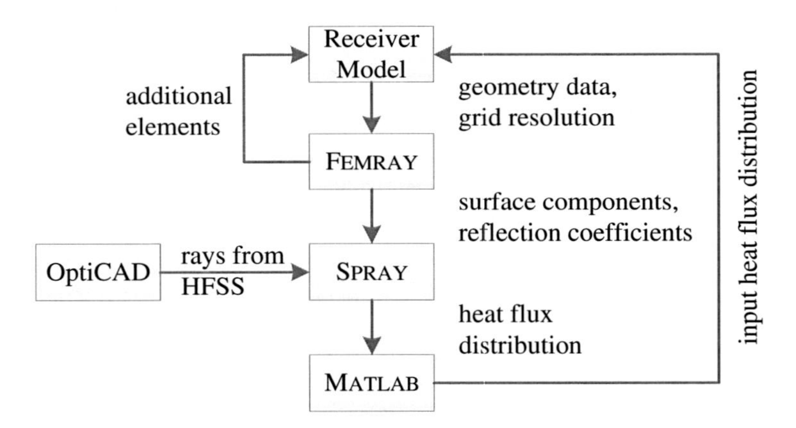

Figure 4.6: Schematic of the procedure to calculate the incoming heat flux distribution and optical losses.

To calculate the incoming heat flux distribution on the receiver the ray-tracing tool SPRAY is utilized. It reads in an output file generated by FEMRAY, that contains the geometry data from the model. In the ray-tracing method a huge number of solar rays are generated and traced while they interact with all components in the examined system. SPRAY is based on the code MIRVAL, originally developed by Sandia National Laboratories for the evaluation of the optical performance of solar tower systems [13]. It has been extensively modified for more flexibility, better performance and espescially for the integration into FEM simulations.

In order to reflect the experimental conditions as realistic as possible the High Flux Solar Simulator (HFSS) providing the inlet power in the experiments is modelled as well. However, the rays are not generated by SPRAY but by the commercially available ray-tracing code OptiCAD® since the HFSS has been already implemented there. Combining the given rays for each lamp from OptiCAD® and the geometry data from FEMRAY a heat flux distribution on all cavity surfaces is generated in SPRAY. In a last step it is converted into an ANSYS macro again and is applied as the boundary condition $\dot{q}_{in}$ from (4.7).

As the implemented model of the HFSS does not exactly correspond to its real arrangement in the experiments, parameters such as lamp position and orientation are adjusted by comparing simulated to measured heat flux distributions. Exemplary flux maps in the aperture plane for measurement and simulation are presented in Figure 4.7. Since the measured flux maps are only of qualitative nature, the heat fluxes in Figure 4.7c and 4.7d are non-dimensionalized by an arbitrary chosen value of $800\,\mathrm{kW\,m^{-2}}$. Good agreement is found indicating the accurate modelling of the incoming heat flux distribution by SPRAY.

4.6 Grid study and validation

A grid study and first validations of the model are conducted for simplified cases. The mesh size basically depends on the number of fluid lines and their revolutions. Arbitrary chosen grid resolutions are investigated and listed in Table 4.1.

For a first validation each heat transfer mechanism is investi-

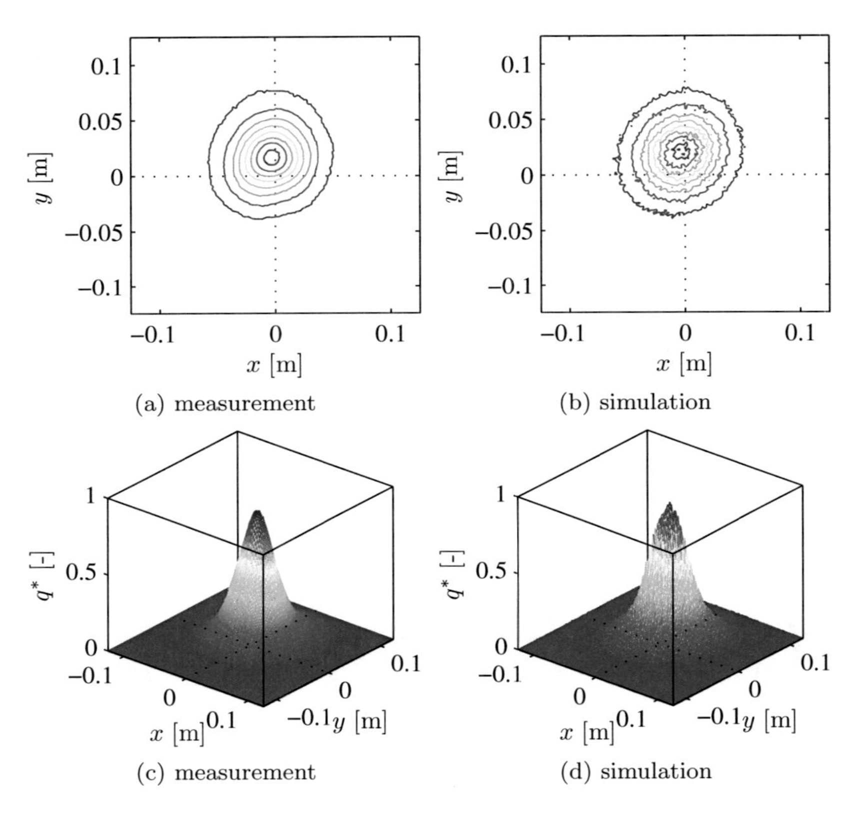

Figure 4.7: Comparison between measurement and simulation of an exemplary heat flux distribution in the aperture plane of one lamp.

Table 4.1: Considered mesh resolutions for grid study and model validation.

Grid	rev	fl	N_E
1	1/2	20	1507
2	1	60	8227
3	2	60	15427
4	3	60	22627
5	5	120	144907

gated in detail. The input parameters for the different cases are summarized in Table 4.2.

Table 4.2: Input parameters for different validation cases.

	T_{in} [°C]	$\dot{m}$ [kg/s]	α [°]	c_p [J/kg/K]	ε [-]	$\dot{q}_{in}$ [W/m^2]
A	25	0.01	90	1000	-	1×10^5
B	500	100	90	1000	0.9	1×10^{-5}
C	25	0.01	90	1000	0.9	1×10^5

In case A the accurate heat transfer to the particles is examined. A homogeneous input heat flux of $\dot{q}_{in} = 1 \times 10^5\,\mathrm{W\,m^{-2}}$ and a low inlet mass flow rate of $\dot{m} = 0.01\,\mathrm{kg\,s^{-1}}$ are applied. Radiation, conduction and convection losses are ignored. The heat capacity of the fluid elements is set temporarily to a constant value of $c_p = 1000\,\mathrm{J\,kg^{-1}\,K^{-1}}$. The heat flow rate $\dot{Q}_{abs}$ absorbed by the fluid lines is compared to the analytical solution according to $\dot{Q} = \dot{m}c_p dT$.

In case B the accuracy of radiation modelling is investigated. The application of $\dot{q}_{in}$ is neglected while a very high mass flow rate is assigned in order to maintain a constant wall temperature. The radiated heat flow rate is compared to the analytical solution calculated with the Stefan-Boltzmann law $\dot{Q} = \varepsilon\sigma A T^4$.

Case C is devoted to the investigation of the energy balance Θ in the simulation, whereas $\Theta = (\dot{Q}_{in} - \dot{Q}_{abs} - \dot{Q}_{rad})/\dot{Q}_{in}$. Radiation calculation is considered and a homogeneous heat flux distribution along with a low mass flow rate are applied.

To quantify the accuracy of the computational method the relative deviation

$$d\Psi^* = |1 - \Psi_{sim}/\Psi_{an}| \tag{4.16}$$

is defined, that relates the numerical result Ψ_{sim} to the analytical solution Ψ_{an}. Ψ represents one of the considered variables, $\dot{Q}_{abs}$, T_o or $\dot{Q}_{rad}$. Relative errors of test cases A and B as well as the energy balance in case C are plotted in Figure 4.8 for different grid sizes (number of elements N_E).

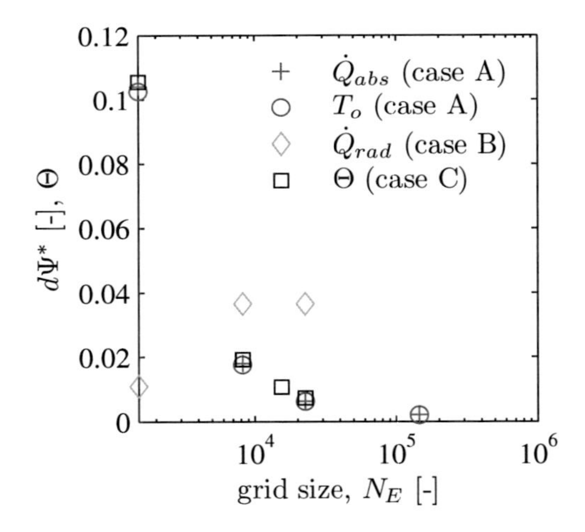

Figure 4.8: Validation cases: Heat absorption by particles (case A), radiation (case B) and energy balance calculations (case C).

It is clearly revealed that the simulation accuracy increases with the number of elements. It appears that grid number 3 with about 150000 elements for the cavity geometry is already sufficiently fine enough as the difference between simulation and analytical solution lies within 1 %. The investigation of the energy balance shows a similar trend. For subsequent simulations, where boundary conditions are chosen according to the experiments, grid sizes with > 150000 elements are utilized since they represent a compromise between accuracy and computational time.

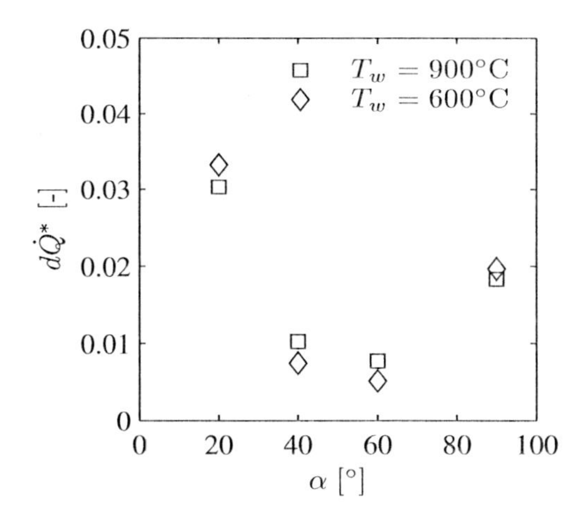

Figure 4.9: Validation of convective loss calculations. The difference between simulation results and analytical data, derived from Clausing's correlation [18], for various α at two different T_w is denoted by $d\dot{Q}^* = |1 - \dot{Q}_{sim}/\dot{Q}_{an}|$. $\dot{m} = 100\,\mathrm{kg\,s^{-1}}$, $N_{rev} = 2$, $N_{fl} = 60$

To evaluate the convection loss calculation, different cases are conducted by varying the inclination angle and mean wall temperature and are compared to results calculated with Clausing's formula [18]. Input power, radiation and conduction are neglected. As indicated in Figure 4.9 the discrepancy between both results is at most about 3 % at $\alpha = 20°$. Considering the approximative character of the correlation itself the achieved accuracy lies within an acceptable range. Note, that Clausing's correlation is actually not valid for $\alpha = 90°$ since he assumed convection losses to be zero for face-down receivers. His assumption however, is not correct as revealed in several experimental investigations [71], [76]. To be still able to use his formula convective losses for $\alpha = 89.9°$ are calculated instead. Since convection at $\alpha = 90°$ plays a minor role in the overall thermal loss evaluation this approximation is sufficiently accurate.

Due to rotation the temperature distribution in a small scale CentRec becomes circumferentially homogeneous at steady-state, independent on the incoming heat flux distribution. As the helical fluid line trajectory represents the receiver rotation, the number of revolutions N_{rev} must be high enough to reflect the homogenizing effect of rotation. This number does not have to compulsorily correspond to the actual particle revolution in the cavity as only steady-state simulations are conducted.

The dependence of the temperature distribution in the fluid lines on N_{rev} is shown in Figure 4.10. All thermal loss calculations are neglected and only the heat input into the particles by an inhomogeneous heat flux distribution is considered. For one third of the cavity a heat flux of $\dot{q}_{in} = 50\,\mathrm{kW\,m^{-2}}$ is applied whereas for the other two thirds a heat flux of $\dot{q}_{in} = 100\,\mathrm{kW\,m^{-2}}$ is applied. The number of fluid lines is considered to be constant at $N_{fl} = 60$.

As assumed, the deviation between the temperature distribution of each fluid line diminishes with increasing revolution number leading to enhanced homogeneity of the particle outlet temperature at $z/L = 0$. As the influence of n_{rev} on the temperature distribution decreases for higher n_{rev} a revolution number between 3 and 4 seems to be sufficient to accurately model the receiver rotation. However, this result might be only applied for

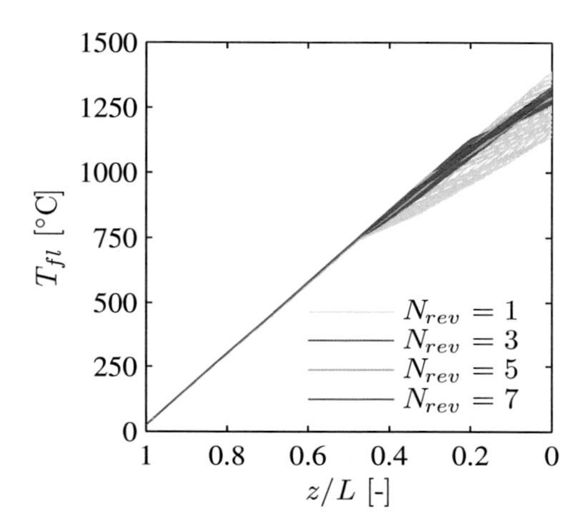

Figure 4.10: Temperature distribution of each fluid line dependent on number of revolutions. The longitudinal axis of the cavity is non-dimensionalized by the cavity length L. The receiver outlet is at $z/L = 0$. $T_{in} = 25\,°\text{C}$, $\dot{m} = 0.01\,\text{kg}\,\text{s}^{-1}$, $N_{fl} = 60$

this specific case. It is possible, that for a different incoming heat flux distribution another sufficiently high n_{rev} must be found.

Due to the small scale of the CentRec prototype, the number of particle revolutions necessary for a homogenized $\dot{q}_{in}$ has to be considerably high. The resulting grid size would exceed the currently available computational power. In order to still consider the rotational effect in the numerical model the applied heat flux distribution is modified instead. Since the cavity wall is discretized in a rectangular grid the mean heat flux for each row can easily be calculated. $\dot{q}_{in}$ is therefore homogenized by calculating the mean heat flux of each row and applying this value to all elements in the corresponding row. Each column or each fluid line is then applied with the same heat flux distribution. As the receiver rotation is already considered by $\dot{q}_{in}$ the modelling of the helical particle trajectories would be now unnecessary. Straight fluid lines are therefore implemented instead. Comparison calculations between this modified and the original model for a simplified case did not yield considerable differences in the solution.

For the present work, this modified version of the model is used as the computational effort can be reduced significantly. However, for scale-up designs of CentRec the original model could be of interest again. Due to the increased dimensions of the receiver, the trajectory of the particles might be only a few revolutions until they are already heated up to the desired temperature at the receiver outlet. Consider now an inhomogeneous irradiation, not all particles will pass through the same heat flux regions despite receiver rotation. The resulting temperature distribution is of great interest in the determination of the thermal receiver performance.

Chapter 5

Experiments

The experiments are divided into two main parts: a feasibility study, where the basic functionality of the receiver concept is demonstrated and important features of the particle behavior are investigated, and subsequent high flux tests, where the receiver is irradiated in the solar simulator. The evaluation of the receiver's thermal performance, its receiver efficiency for various operation parameters as well as model validation are of main interest in the latter experiments.

5.1 Feasibility study

In first proof-of-concept tests the basic functioning of the CentRec concept needed to be proven. The experiments are focused on answering two main questions: first, is it possible to generate a thin and optically dense particle layer, that slowly moves downwards towards the aperture? And second, can the particle residence time be controlled by adjusting the rotational speed and the mass flow rate in order to ensure a thin dense layer for all load conditions? The test rig has been therefore modified in such a way that an optical access into the receiver inside is facilitated and the particle movement can be qualitatively recorded by a high-speed camera. The experimental set-up is presented in Figure 5.1. The high-speed camera is placed along the receiver axis. In order to be able to observe as much as possible of the cavity walls, the collecting ring is exchanged by a plexiglas container and three to four spotlights are mounted around the camera illuminating the receiver's insight. High flux irradition is neglected as the main focus lies on the particle behavior.

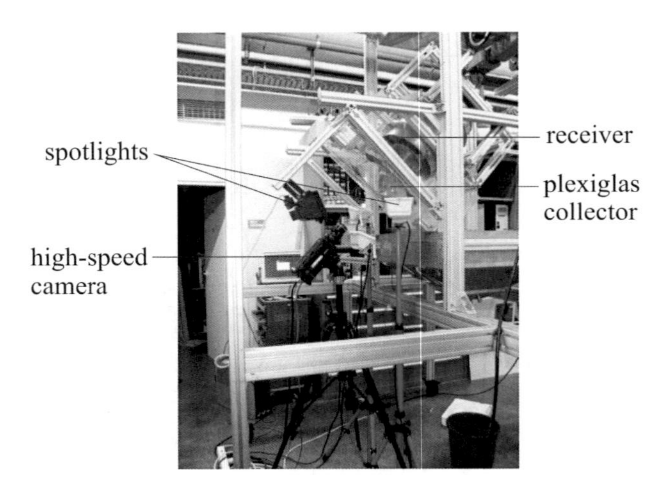

Figure 5.1: Experimental set-up for feasibility study.

For visualization purposes white colored particles are added to build a contrast to the black particles. A typical picture from the camera's point of view is displayed in Figure 5.2a. Due to clarity reasons, the main visible parts of the receiver are presented and named additionally in the CAD sketch in Figure 5.2b. As instruments and measurement methods were limited, just qualitative results of the particle behavior are gained at the moment. However, for a first feasibility test and the answering of the first question, the present recordings are sufficiently detailed.

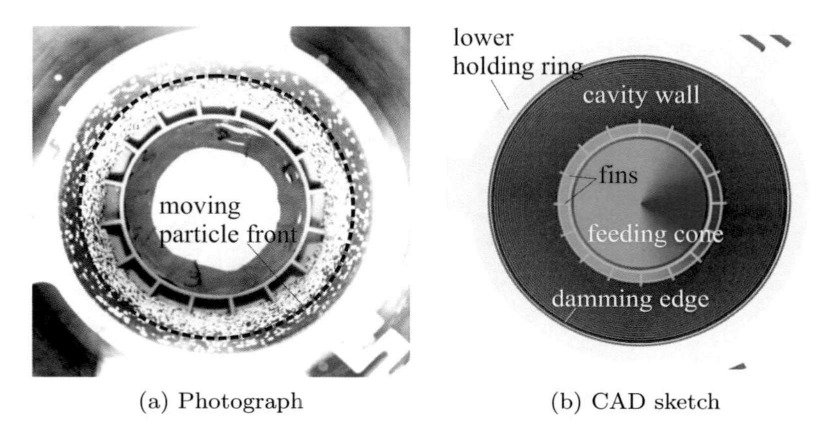

(a) Photograph

(b) CAD sketch

Figure 5.2: Typical picture taken by high-speed camera (a). CAD sketch of same picture for notation purposes of parts (b).

5.1.1 Dense particle film

For the development of an optically dense particle layer, the particles must stick to the receiver wall on the one hand and on the other hand still be able to move slowly downwards. The wall roughness is here of crucial importance. If it is not high enough, particles will not stick to the wall and no dense layer can be formed. Preliminary experiments have revealed, that a wall roughness of at least one particle radius seems to be adequate.

As the wall surface of the Inconel tube exhibit an insufficient roughness and restricted machining possibilities due to the hardness of the material, a 0.8 mm width damming edge has been fabricated at the tube outlet. Experiments have demonstrated that this edge causes an accumulation of particles which initially enter the receiver in the start-up process. They form a first stationary base layer providing the necessary wall roughness for following particles. A second particle layer is built, that slowly moves over the base layer with a speed defined by the receiver rotation rate. A schematic of the described particle behavior is sketched in Figure 5.3.

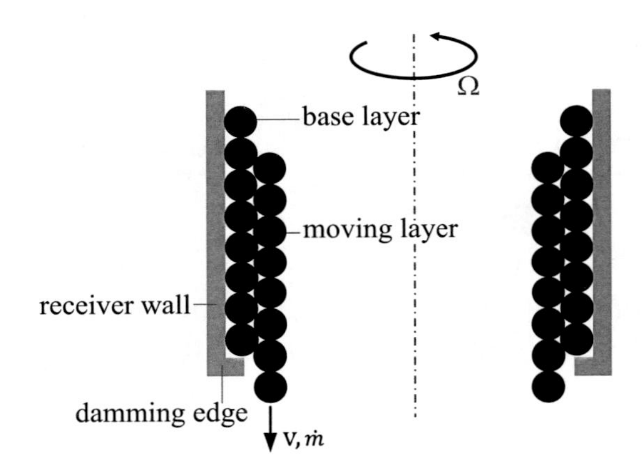

Figure 5.3: Schematic of the particle layer development caused by the damming edge.

With this base layer building the required wall roughness, sufficient friction is assured and the first premise for a dense particle layer is given. For a given mass flow rate and the corresponding rotation rate, a slowly downwards moving dense particle film can be indeed achieved, which is substantiated by recordings of the high-speed camera in Figure 5.4. The film behavior in different time steps is presented. White colored particles move over the base layer of black particles as a nearly uniform circular front. Gaps within the moving layer can be observed at no time. As the film reaches the outlet the cavity surface is fully covered with white colored particles (Figure 5.4f). The first question is hence successfully answered.

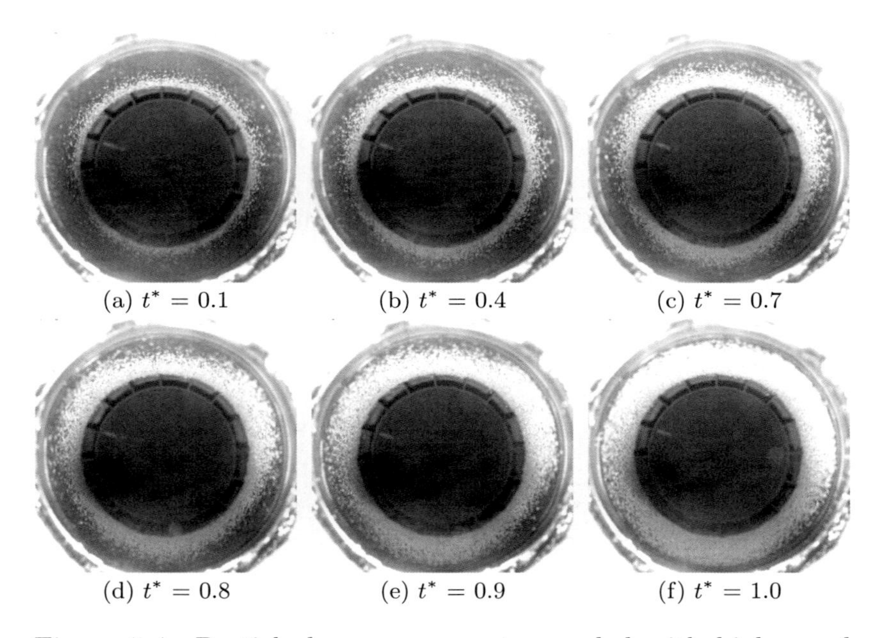

Figure 5.4: Particle layer movement recorded with high-speed camera. t^* denotes the recording time t non-dimensionalized by the particle retention time t_{ret}.

5.1.2 Particle movement

One can assume that there is one rotation speed matching one mass flow rate, where a thin and dense particle film is formed, that continuously moves towards the receiver outlet. However, the exact combination of both is hard to find. Additional effects, which are not entirely avoided in experiments, like eccentricity, unbalance and vibration of the holding frame structure can have a significant impact on the particle movement. In fact, mass flow rate measurements in initial experiments have revealed that the axial movement of the particle film does not happen in a homogeneous and constant manner but more in a periodical way. Such a behavior is exemplarily displayed in Figure 5.5 for a nominal mass flow rate of $\dot{m}_{nom} = 6\,\mathrm{g\,s^{-1}}$. The measured outcoming mass flow rate oscillates between low and high mass flow rates, ranging from 1 to $13\,\mathrm{g\,s^{-1}}$. It is assumed that in times where $\dot{m} < \dot{m}_{nom}$ (denoted by the yellow coloured area) a particle accumulation occurs in the receiver resulting in a decreased mass flow rate. However, after a critical accumulated mass has been reached, excess particles drain off the receiver resulting in a significant increase of the measured mass flow rate, such that $\dot{m} > \dot{m}_{nom}$ (blue coloured area).

More detailed investigations lead to one possible explanation that is sketched in Figure 5.6a. For unfavorable combinations of rotation speed and mass flow rate, particles in the receiver do not move as fast as new particles following from the feeding cone. They are concentrated around the receiver inlet up to a critical angle γ. When exceeding this angle, all accumulated particles slip off and leave the receiver as a flood. This happens periodically in certain time intervals.

To underline this theory, recordings from the high-speed camera are presented. Figure 5.6b is focused on the receiver inlet while Figure 5.6c is focused on the outlet. The fins of the feeding cone give a good estimation of the particle layer thickness around

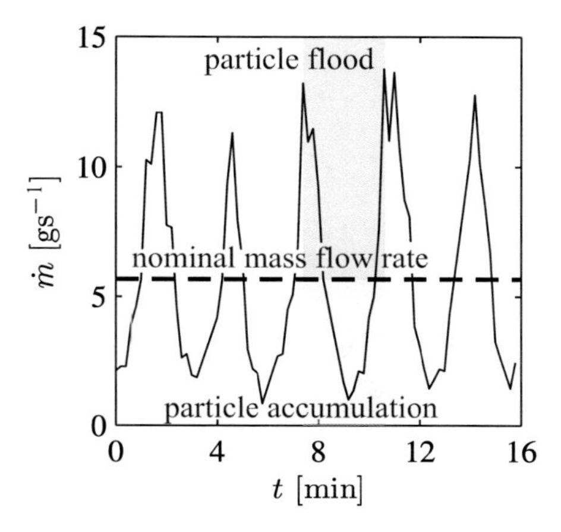

Figure 5.5: Exemplary mass flow rate measurement without intended receiver vibration. Dashed line denotes the nominal mass flow rate of $\dot{m}_{nom} = 6\,\mathrm{g\,s^{-1}}$.

the inlet region. The less visible the fins are, the thicker the film is. (Compare to Figure 5.2a for example, where a thin layer is presented.) Considering Figure 5.6b just a small part of the fins is apparent indicating a film which is most likely several particle layers thick. Examine now the corresponding outlet region, the damming edge is clearly still visible. As this edge is smaller than one particle diameter and the film above seems not to stick out significantly, it can be assumed that the film here is quite thin, consisting of just one or two particle layers. However, these are just qualitative observations and do not give a final proof of this theory which must be left to subsequent research.

For the present experiments, an intended receiver vibration is hence implemented in order to ensure a homogeneous and stable particle flow. An additional fourth "vibration wheel", prepared with small notches on the tread, is installed on the stationary holding frame next to the three hard rubber rollers for the ra-

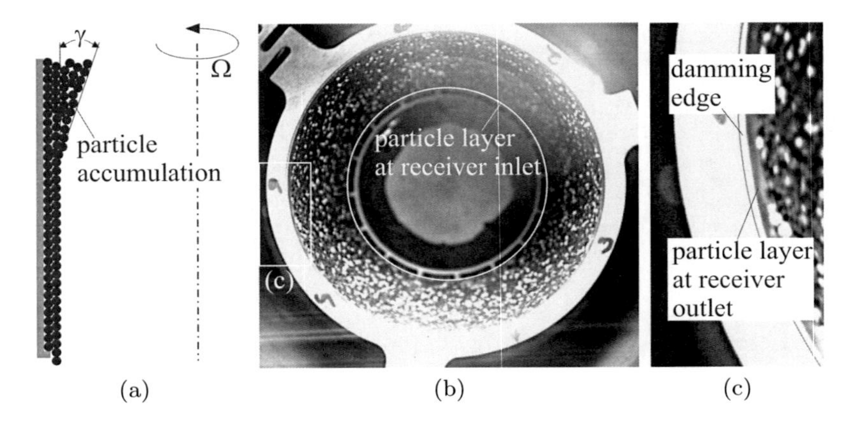

Figure 5.6: Particle layers at receiver inlet and outlet without receiver vibration.

dial centering of the receiver (Chapter 3.1.1). Depending on the pressing force of the vibration wheel on the guide ring various vibration intensities can be set.

5.1.3 Critical rotation speed and retention time

It has been successfully demonstrated that a dense particle film can indeed be created. To get an impression of the controllability of the film and to address the second question, experiments are conducted where the influence of mass flow rate and rotation rate is evaluated. For comparison and scalability reasons two characteristic variables are introduced. The mass flow rate is related to the estimated particle mass of one dense layer m_{1p} with

$$\dot{m}^* = \frac{\dot{m}}{m_{1p}} = \frac{\dot{m}}{2\pi R_{cav} L \delta_{1p} \rho_b}. \tag{5.1}$$

m_{1p} is dependent on the receiver geometry with R_{cav} as the radius of the inner cavity wall and L as the cavity length. The film thickness of one particle layer is denoted by δ_{1p} and ρ_b represents the bulk density. $\dot{m}^*$ can be seen as the inverse of the required minimum retention time for the development of one dense particle layer. If the particles move faster through the receiver in less than this time, no dense film can evolve.

The second characteristic variable is the non-dimensionalized rotation speed Ω^*, relating centrifugal to gravitational acceleration,

$$\Omega^* = \frac{\mu R_{cav}}{g} \Omega^2. \tag{5.2}$$

The friction coefficient μ accounts not only for wall roughness, but also for vibrational effects as intendedly applied in the current set-up. However, μ is unknown in the present work and will be omitted for now. For the particles to be pressed against the cavity wall, Ω^* must be > 1 as gravitational force has to be overcome by centrifugal force. There exists also an upper limit for Ω^* when no more particle movement is possible.

The rotation rate determines the axial velocity of the particle film. If it is not high enough for a specific mass flow rate, the particle film would move faster through the cavity than following particles. For each mass flow rate there exists one critical minimum rotation speed Ω_{crit} for which the development of a dense particle film is possible. Figure 5.7a depicts the corresponding Ω^*_{crit} for three different mass flow rates at a receiver inclination of $\alpha = 45°$. The relative mass flow rates $\dot{m}^*$ correspond to absolute mass flow rates of 2, 6 and 12 $\mathrm{g\,s^{-1}}$.

The critical rotation speed is experimentally determined by visual observation of the particle film. For a given mass flow rate, an arbitrary rotation rate is chosen to generate a stable and continuously moving particle film. After steady-state has been reached, the rotation speed is gradually decreased until the point, where a thin, stable film is barely possible. Further decreasing would lead to a breakdown of the film. As expected, the lower the mass flow rate, the higher is Ω_{crit} as the particle retention time in the receiver must be extended in order to obtain the cavity surface to be fully covered by a dense particle film. As Ω is considered in the square for the centrifugal acceleration, the altering range of the critical rotation speed is quite small. While the mass flow rate is reduced by one half (from 12 to 6 g s^{-1}), Ω_{crit} is only increased by about 1.5 % (from 157.2 to 159.6 rpm).

Since CentRec can be operated in different inclination angles, the dependence of Ω_{crit} on α is also investigated. Results are shown in Figure 5.7a, where the increase of Ω_{crit} with higher tilt angle is clearly apparent. This observation matches the expectation as gravitational force is factorised in radial and axial direction for inclined receivers. The fraction in axial direction is therefore smaller but grows with increasing α. For $\alpha = 90°$ no radial part of g is present and the centrifugal force must overcome the entire gravitation in axial direction.

However, the influence of α on Ω_{crit} is also strongly dependent on μ, especially with regard to vibrational effects. Experiments have revealed that the receiver vibration for $\alpha = 90°$ has to be increased in order to ensure a steady particle flow. The periodic behavior of the mass flow rate, discussed above, seems to be stronger in the face-down position. The particle accumulation around the receiver inlet is favoured as the radial force is always constant over the entire circumference. In contrast, this force is altered in an inclined receiver as the radial part of gravitation points one time in the same and another time in the opposite direction of the centrifugal force. Particles hence experience a varying radial force resulting in a weaker accumulation. The receiver vibration in 90° position needs thus to be increased leading in turn to higher Ω_{crit} for the generation of a dense particle film.

Figure 5.7b presents the particle retention time t_{ret} depending on the rotation rate for two different mass flow rates. t_{ret} is estimated by measuring the time between two high-speed camera recordings, where white colored particles enter the receiver

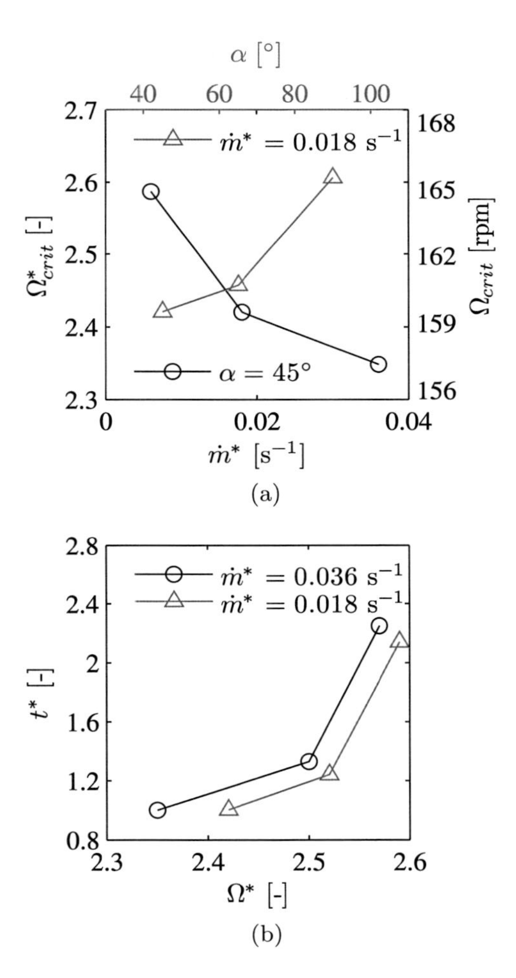

Figure 5.7: Critical rotation speed and particle retention time for various operation parameters, such as mass flow rate and inclination angle.

and where the cavity wall is fully covered by them. As these investigations are more of a qualitative nature, retention times are related to t_{ret} for the lowest rotation rate in order to get an impression of their relative change to each other.

As expected, t_{ret} increases with higher Ω as centrifugal forces increase. The quadratic dependence of centrifugal forces on the rotation rate is clearly indicated as t_{ret} exhibits a steep increase from the second to the third considered Ω. This trend is observed for both mass flow rates. Matching the findings in Figure 5.7a for higher mass flow rates, Ω is lower for the same retention time.

Examining corresponding recordings of the high-speed camera, a thicker particle film is observed for higher Ω. This can be explained by increased centrifugal forces on the particles leading to a decrease of their axial velocity. As the mass flow rate maintains constant a growth of the film thickness in radial direction is the consequence. However, due to the qualitative nature of the measurement procedure the number of actually moving particle layers could not be finally evaluated. The identification of this number is essential as discussed in the next section.

The presented results have clearly revealed, that a dense particle film indeed can be achieved for all considered mass flow rates by adjusting Ω. Moreover, by setting an appropriate rotation rate the particle retention time can be regulated and the feasibility of the proposed receiver concept could be successfully demonstrated.

5.1.4 Discussion

Since the present work is focused on the demonstration of the feasibility of the proposed receiver concept and espescially its thermal performance, the investigations regarding the particle behaviour are of a rather qualitative nature. For subsequent research a more detailed look should thus be taken into a series of interesting problems.

One important question which could not be sufficiently addressed so far, is the number of actually moving particle layers. It is expected that for an optimum thermal efficiency, the moving particle film should be relatively thin, consisting of just one or two layers in order to avoid high temperature gradients within the moving layers. If substantially more layers would move the risk exists, that only particles of the innermost layer would be heated directly while particles of the outermost layer would remain relatively cold. A high temperature gradient between inner- and outermost layer would occur resulting in a mixed outlet temperature that is significantly below the peak film temperature. Due to the thicker film layer the particle retention time is increased, and considerably higher particle temperatures of the innermost layer can evolve resulting in increased radiation losses and therefore lower receiver efficiencies.

The sensitivity of the particle behaviour to external effects like unbalances, eccentricity and vibration, leading to an inhomogeneously moving particle film along the circumference, has been encountered in several experiments. The influence of these external effects is not negligible and should be therefore investigated and quantified in more detail. In this context, one interesting question would be, if it is possible to generate a thick stationary particle layer that can balance ovalities of the receiver walls leading only the innermost particle layer to move uniformly along the circumference despite the actual form of the cavity. This is favourable espescially for the scale-up of receivers as production tolerances could be increased. In the course of these investigations questions like the minimum achievable thickness of the stationary layer and the maximum possible equilibration of ovalities and eccentricities should be addressed.

Another major aspect is the influence of the receiver vibration on the particle behaviour. The vibration wheel has been primarily installed in order to ensure a steady particle flow without closer consideration. In subsequent research the experimental set-up should be modified in such a way that vibrational effects can be investigated in more detail with regard to amplitude, frequency and position. Moreover, efforts could be undertaken in order to quantify the friction coefficient μ and consider it in describing the particle behaviour.

5.2 High flux tests

In order to examine the thermal performance of the receiver and its efficiency as well as providing a data basis for model validation, high flux experiments are conducted in the solar simulator. Two relevant receiver inclinations are investigated in detail, $\alpha = 45°$ and $\alpha = 90°$.

The general experimental procedure starts with the generation of a stable and homogeneously moving particle film. When the particle movement reaches steady-state, a defined irradiation is applied and the receiver is gradually heated up. By operating the TMR particle outlet temperatures are measured when the monitored receiver wall temperatures indicate a steady state. The measurement period of the TMR lies within three to four minutes. Particle mass flow and rotation rates are evaluated continuously during the entire experiment.

5.2.1 Receiver inclination 45°

As $\alpha = 45°$ is one inclination angle of interest according to preliminary HFLCAL simulations in Chapter 1.2.2, the receiver performance is investigated at this particular inclination for different mass flow rates and input power. A summary of the results is presented in Table 5.1. The average particle outlet temperature $\overline{T}_o = T_o$ is calculated according to (3.8) in Chapter 3.3.1. Uncertainties of the measurements are listed in detail in Appendix B.

Table 5.1: Experimental results at $\alpha = 45°$.

	$\dot{q}_{in} = 265\,\mathrm{kW\,m^{-2}}$	$\dot{q}_{in} = 370\,\mathrm{kW\,m^{-2}}$
$\dot{m}_{nom}$ [$\mathrm{g\,s^{-1}}$]	T_o [°C]	
9.5	385	487
6.0	539	688
4.0	669	801
3.0	745	878

For $\dot{m}_{nom} = 9.5\,\mathrm{g\,s^{-1}}$, the interpolated distribution of measured wall temperatures is shown for two input heat fluxes in Figure 5.8a and 5.8b. While the x-axis denotes the circumferential direction φ, the y-axis denotes the non-dimensionalized axial axis of the receiver, where $z/L = 1$ corresponds to the receiver inlet and $z/L = 0$ to the outlet. The black crosses in Figure 5.8a exemplarily represent the measuring thermocouple positions directly behind the receiver wall as described in Chapter 3.2.1.

The temperature increase from inlet to outlet clearly demonstrates the heating process of the moving particles by the incoming radiation. The temperature distribution in circumferential direction however is obviously quite uniform indicating a homogeneously distributed and moving particle layer.

The more qualitative examinations of the interpolated temperatures are substantiated by the lower diagrams in Figure 5.8, where the actual temperatures directly measured by the thermocouples are displayed. The temperature profile from inlet to outlet reveals a parabolic shape, where the most intense heating of the particles occurs apparently in the receiver outlet region. This is indicated by an increasing temperature gradient with decreasing z/L.

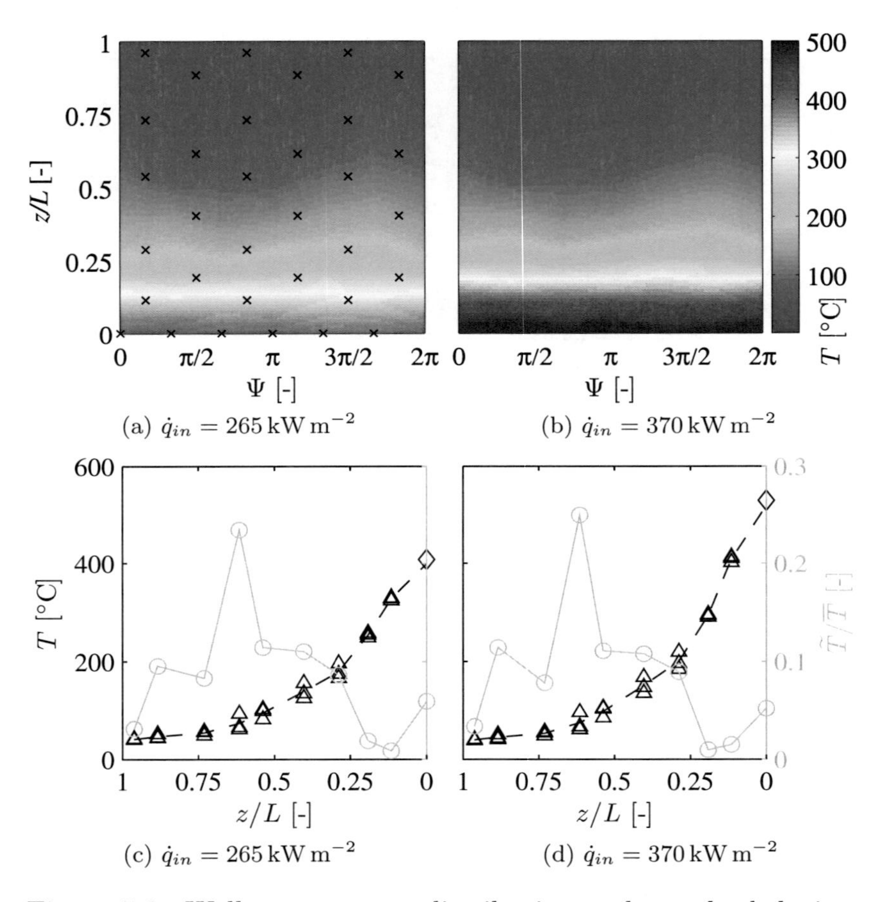

(a) $\dot{q}_{in} = 265\,\mathrm{kW\,m^{-2}}$

(b) $\dot{q}_{in} = 370\,\mathrm{kW\,m^{-2}}$

(c) $\dot{q}_{in} = 265\,\mathrm{kW\,m^{-2}}$

(d) $\dot{q}_{in} = 370\,\mathrm{kW\,m^{-2}}$

Figure 5.8: Wall temperature distribution and standard deviation for two input heat fluxes with $\dot{m}_{nom} = 9.5\,\mathrm{g\,s^{-1}}$. Discrete thermocouple positions are exemplarily denoted by crosses in (a). ($\triangle$) $T_{w,TC}$, ($\diamond$) $T_{o,TC}$, (-$\ominus$-) $\tilde{T}/\overline{T}$

As the thermocouples are distributed in such a way, that there are always three measuring TCs at the same axial but at different circumferential positions, the standard deviation (SD) of these three measurements is derived in order to gain an impression of the quantitative homogeneity of the temperatures. The SD of the temperature measurements, $\tilde{T}$, related to the mean $\overline{T}$ is represented by the grey circles in Figure 5.8c and Figure 5.8d. Apparently, the highest temperature deviation of almost 25 % occurs at around $z/L = 0.6$, about one third after the receiver inlet. This peak could indicate a slightly inhomogeneous particle movement, but as $\tilde{T}/\overline{T} = 25\,\%$ corresponds to an absolute deviation of about 22 °C at an receiver region, where the mean wall temperature is about $\overline{T}_w = 90\,°C$, this difference is negligible. The deviation decreases as particles move further towards the outlet to about less than 1 % around $z/L = 0.2$. This behavior seems not to vary much for different input power levels.

The particle outlet temperature is directly measured by six thermocouples as described in Chapter 3.2.1. As expected, T_o increases with higher input power from 385 °C to 487 °C. The SD related to the mean is about 8 % for both power levels corresponding to absolute deviations of around 30 to 40 °C for T_o, which lies in an acceptable range.

The measured and interpolated wall temperature distribution for the remaining parameters from Table 5.1 are brought out in the following Figures 5.9, 5.10 and 5.11. In general, temperature characteristics similar to those for the case with $\dot{m}_{nom} = 9.5\,\mathrm{g\,s^{-1}}$ are observed. However, the parabolic shape of the temperature increase in axial direction approximates more a linear progression as the overall temperature level is raised with decreasing mass flow rate.

Moreover, the circumferential homogeneity of the wall temperatures grows with lower mass flow rates which is revealed by a decreasing standard deviation. However, that must not directly

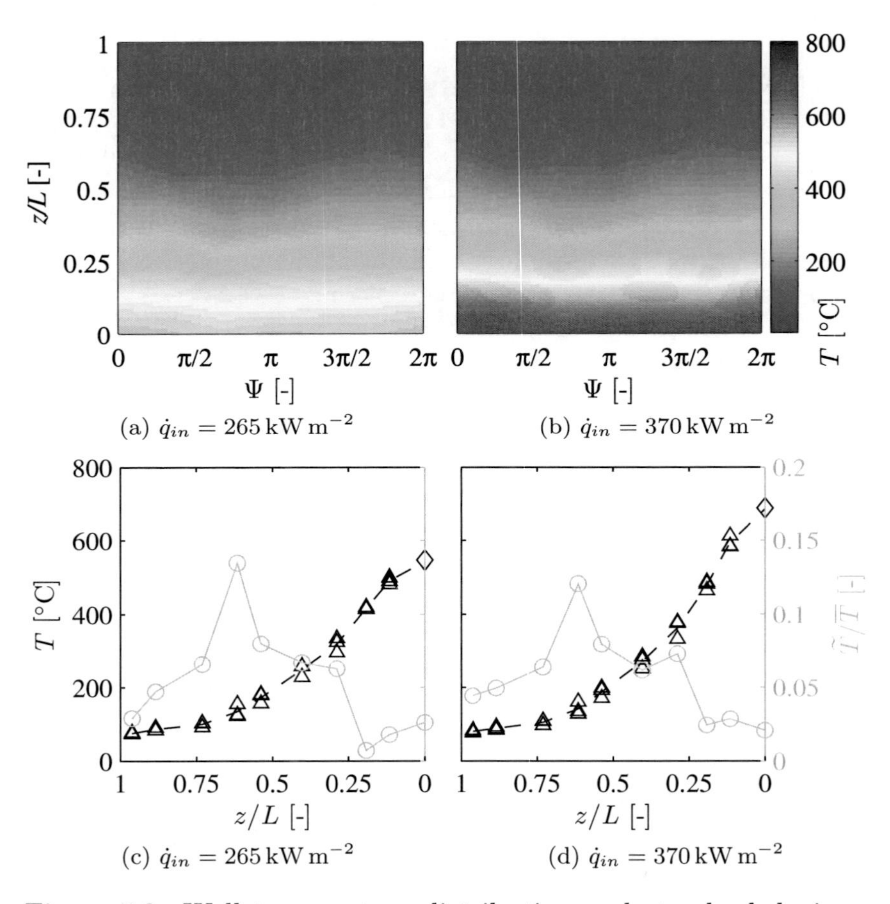

Figure 5.9: Wall temperature distribution and standard deviation for two input heat fluxes with $\dot{m}_{nom} = 6\,\mathrm{g\,s^{-1}}$. (△) $T_{w,TC}$, (◊) $T_{o,TC}$, (-⊖-) $\tilde{T}/\overline{T}$

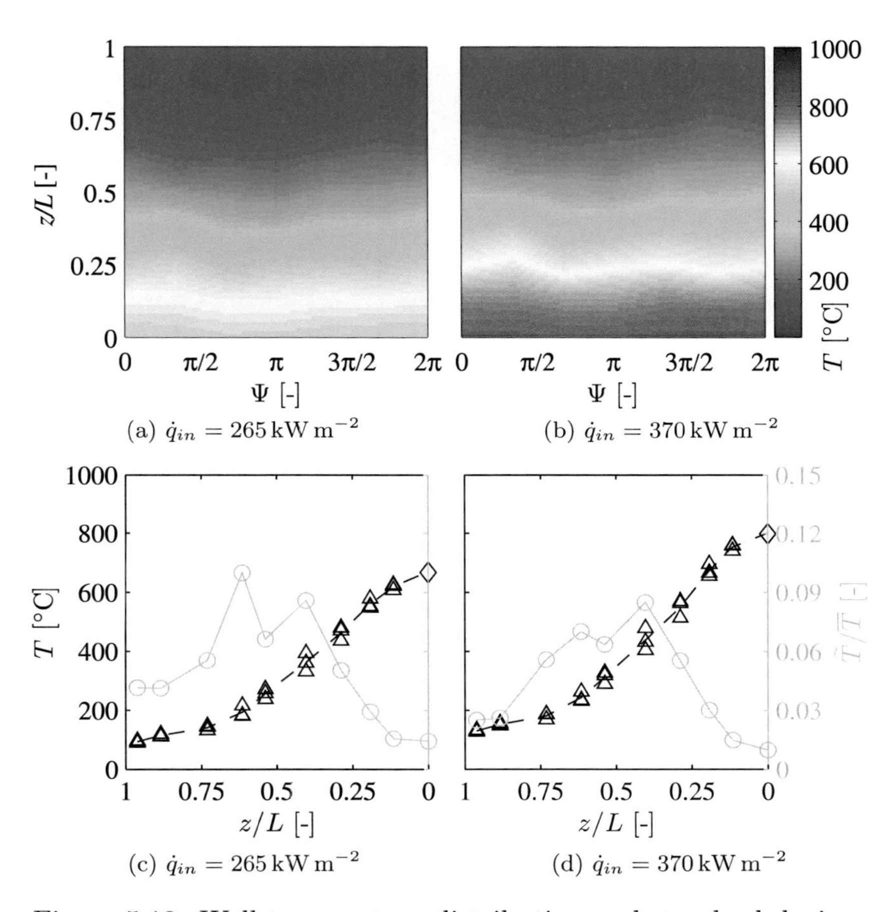

(a) $\dot{q}_{in} = 265\,\mathrm{kW\,m^{-2}}$ (b) $\dot{q}_{in} = 370\,\mathrm{kW\,m^{-2}}$

(c) $\dot{q}_{in} = 265\,\mathrm{kW\,m^{-2}}$ (d) $\dot{q}_{in} = 370\,\mathrm{kW\,m^{-2}}$

Figure 5.10: Wall temperature distribution and standard deviation for two input heat fluxes with $\dot{m}_{nom} = 4\,\mathrm{g\,s^{-1}}$. ($\triangle$) $T_{w,TC}$, ($\Diamond$) $T_{o,TC}$, ($\ominus$) $\tilde{T}/\overline{T}$

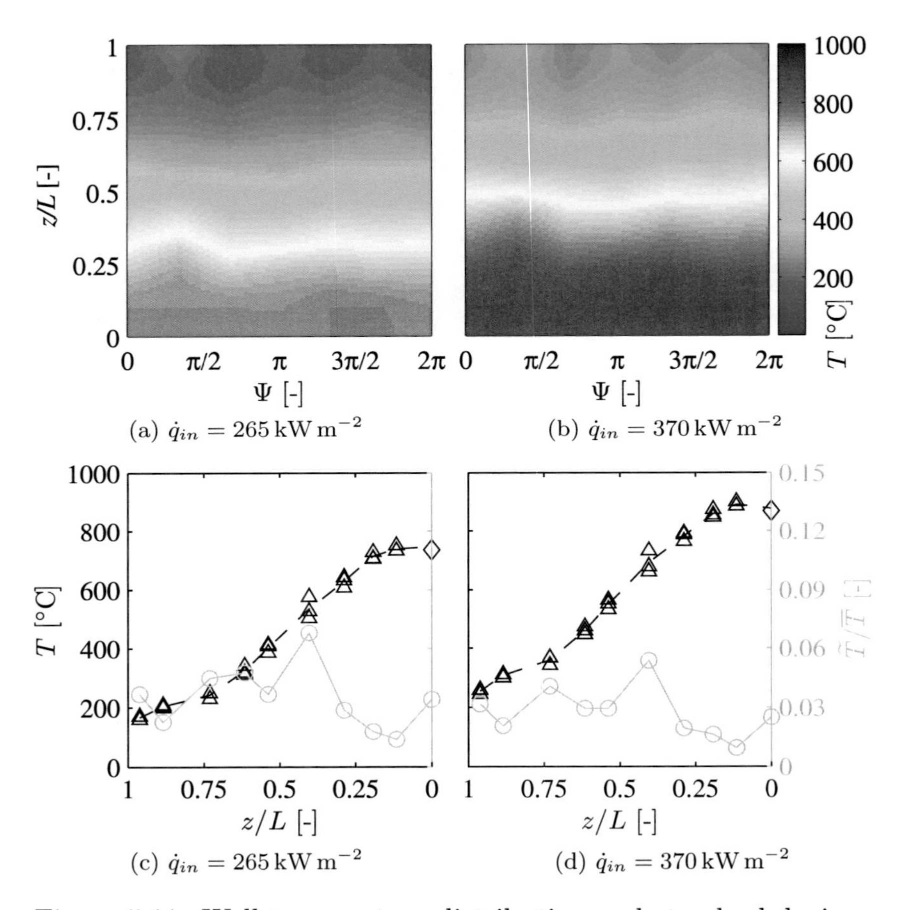

(a) $\dot{q}_{in} = 265\,\mathrm{kW\,m^{-2}}$ (b) $\dot{q}_{in} = 370\,\mathrm{kW\,m^{-2}}$

(c) $\dot{q}_{in} = 265\,\mathrm{kW\,m^{-2}}$ (d) $\dot{q}_{in} = 370\,\mathrm{kW\,m^{-2}}$

Figure 5.11: Wall temperature distribution and standard deviation for two input heat fluxes with $\dot{m}_{nom} = 3\,\mathrm{g\,s^{-1}}$. ($\triangle$) $T_{w,TC}$, ($\Diamond$) $T_{o,TC}$, (-⊖-) $\tilde{T}/\overline{T}$

mean, that the particle movement is more homogenized with smaller mass flow rates. It can be rather assumed, that since the overall temperature level of the receiver wall is enhanced, thermal balancing mainly due to radiation prevails leading to a more uniform temperature distribution.

Considering Figure 5.11, one of the main objectives of the work could be successfully demonstrated. The target particle outlet temperature of 900 °C could be closely achieved for $\dot{m}_{nom} = 3\,\mathrm{g\,s^{-1}}$ and $\dot{q}_{in} = 370\,\mathrm{kW\,m^{-2}}$. Accounting for measurement uncertainties and similar values on IR camera recordings this measured temperature can be seen as a reliable value.

An interesting aspect, which is further observed, is the increase of the first temperature, measured by the thermocouples close to the receiver inlet, up to 250 °C for $\dot{m}_{nom} = 3\,\mathrm{g\,s^{-1}}$ and $\dot{q}_{in} = 370\,\mathrm{kW\,m^{-2}}$. Assuming this temperature to be a reference to the particle inlet temperature and considering the feeding cone temperature of over 330 °C, it can be supposed that the particles might be preheated before they enter the receiver. This preheating process seems to be enhanced with higher temperature levels.

IR camera recordings of the receiver wall for $\dot{m}_{nom} = 4\,\mathrm{g\,s^{-1}}$ and an input heat flux of $\dot{q}_{in} = 370\,\mathrm{kW\,m^{-2}}$ are presented exemplarily in Figure 5.12. As the accurate temperature measurement by an IR camera depends on the emissivity which is not exactly known for the particles, the measurements are only taken for a qualitative comparison. A temperature scale is therefore left out. All six figures, each representing one particular circumferential angle, reveal almost no deviations supporting the above findings of a homogeneous temperature distribution.

In Figure 5.13 temperature profiles along the straight line marked in Figure 5.12a are qualitatively compared to the ones measured by thermocouples from Figure 5.10d. IR measurements are non-dimensionalized by the maximum temperature at position $\varphi = 0$ while TC temperatures are related to the mean outlet

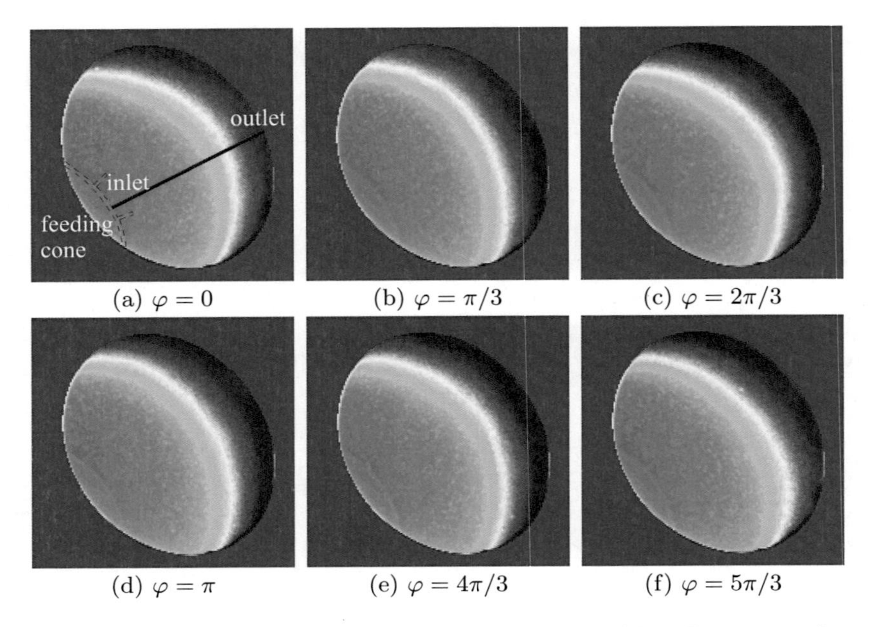

(a) $\varphi = 0$ (b) $\varphi = \pi/3$ (c) $\varphi = 2\pi/3$

(d) $\varphi = \pi$ (e) $\varphi = 4\pi/3$ (f) $\varphi = 5\pi/3$

Figure 5.12: Temperature distribution recorded by IR camera for $\dot{m}_{nom} = 4\,\mathrm{g\,s^{-1}}$ and $\dot{q}_{in} = 370\,\mathrm{kW\,m^{-2}}$. Qualitative comparison of circumferential distribution, therefore no temperature scale is shown.

temperature. Similar distributions for $z/L \leq 0.4$ are clearly visible. The IR recordings for $z/L > 0.4$ can not be considered as correct as the wall temperature here is below the working range of the IR camera from 300 to 1000 °C.

The SD related to the mean of the IR measurements are also presented in Figure 5.13. Maximum deviations of up to $\tilde{T}_{IR}/\overline{T}_{IR} = 4\,\%$ are in the same order of magnitude as for temperatures measured by thermocouples.

Another crucial aspect is the influence of the rotation rate Ω on the particle behavior and thus on the thermal receiver performance. As already demonstrated in Chapter 5.1.3 increasing Ω

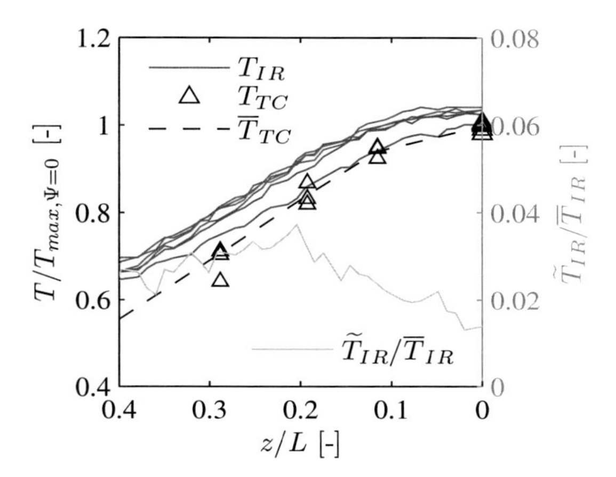

Figure 5.13: Qualitative comparison of temperature distributions measured with the IR camera and thermocouples. The red straight lines denote the temperature distribution along the line displayed in Figure 5.12a for each considered circumferential position shown in Figure 5.12.

leads to a growth of the particle layer thickness along with higher particle retention times. The effect of this phenomenon on the particle outlet temperature is thus investigated for the case with $\dot{m}_{nom} = 4\,\mathrm{g\,s^{-1}}$ and $\dot{q}_{in} = 265\,\mathrm{kW\,m^{-2}}$. Due to the quadratic dependence of centrifugal forces on the rotation speed the possible altering range of Ω is quite small (just 2 to 3 rpm). The lowest rotation speed corresponds to a thin film thickness of 2 to 3 particle layers whereas the highest rotation speed corresponds to a thick film of probably 10 to 12 particle layers.

Measurements of mass flow rate and corresponding wall and outlet temperatures at three different Ω are depicted in Figure 5.14 for a steady state time of 20 minutes each. Peaks in the mass flow rate diagram can be ignored as they are at TMR opening times (denoted by black squares), where particles from chambers are released at once. Obviously, Ω does not affect the measured mass flow rate in a crucial way. Although the amplitude of the mass flow rate fluctuations seem to decrease with higher Ω, the mean value of all three cases is still about $\dot{m}_{nom} = 4\,\mathrm{g\,s^{-1}}$.

The distribution of wall and particle outlet temperature, presented on the right of Figure 5.14, reveal similar characteristics. In the considered time range, the wall temperatures remain nearly constant exhibiting just little deviations while the outlet temperature is measured almost the same for each TMR cycle. It seems that in the investigated range of rotation rates between 172.2 and 174.6 rpm, varying Ω does not significantly influence thermal receiver performances.

The above presented investigations were conducted for a constant and sufficiently high receiver vibration, which is crucial for ensuring a continuous and steady particle movement according to the postulation in Chapter 5.1.2. Figure 5.15 demonstrates the consequences of a deficient vibration especially on the thermal receiver performance for the case with $\dot{m}_{nom} = 6\,\mathrm{g\,s^{-1}}$ and $\dot{q}_{in} = 265\,\mathrm{kW\,m^{-2}}$.

Beginning with the smallest Ω, Figure 5.15a exhibits heavy oscillations of the mass flow rate from about 2 to $14\,\mathrm{g\,s^{-1}}$ and a period of around four minutes. These oscillations indicate irregular particle flow behavior as already observed in the feasibity study (Chapter 5.1.2). Low $\dot{m}$ means that particle layers are built up and accumulated near the receiver inlet, whereas high $\dot{m}$ corresponds to the accumulated layer being destabilized and leaving the receiver as a flood.

This mass flow rate behavior is directly reflected by the wall temperatures which oscillate with a similar period and an amplitude of about 60 to 70 °C (Figure 5.15b). The particle outlet temperature deviates accordingly, depending on the mass flow rate phase at which the TMR is activated. Higher temperatures are measured during a phase with small mass flow rates whereas

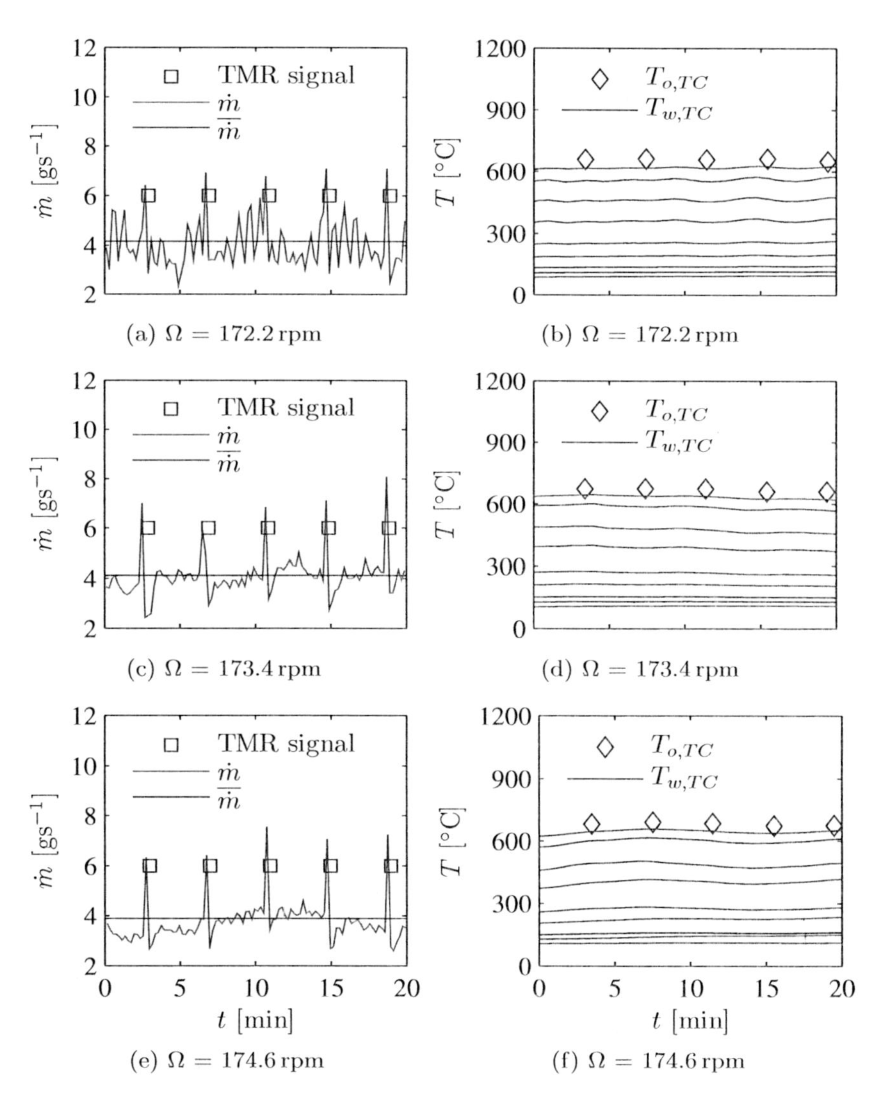

Figure 5.14: Particle temperature and mass flow rate for various Ω with sufficiently high receiver vibration, $\dot{m}_{nom} = 4\,\mathrm{g\,s^{-1}}$, $\dot{q}_{in} = 265\,\mathrm{kW\,m^{-2}}$.

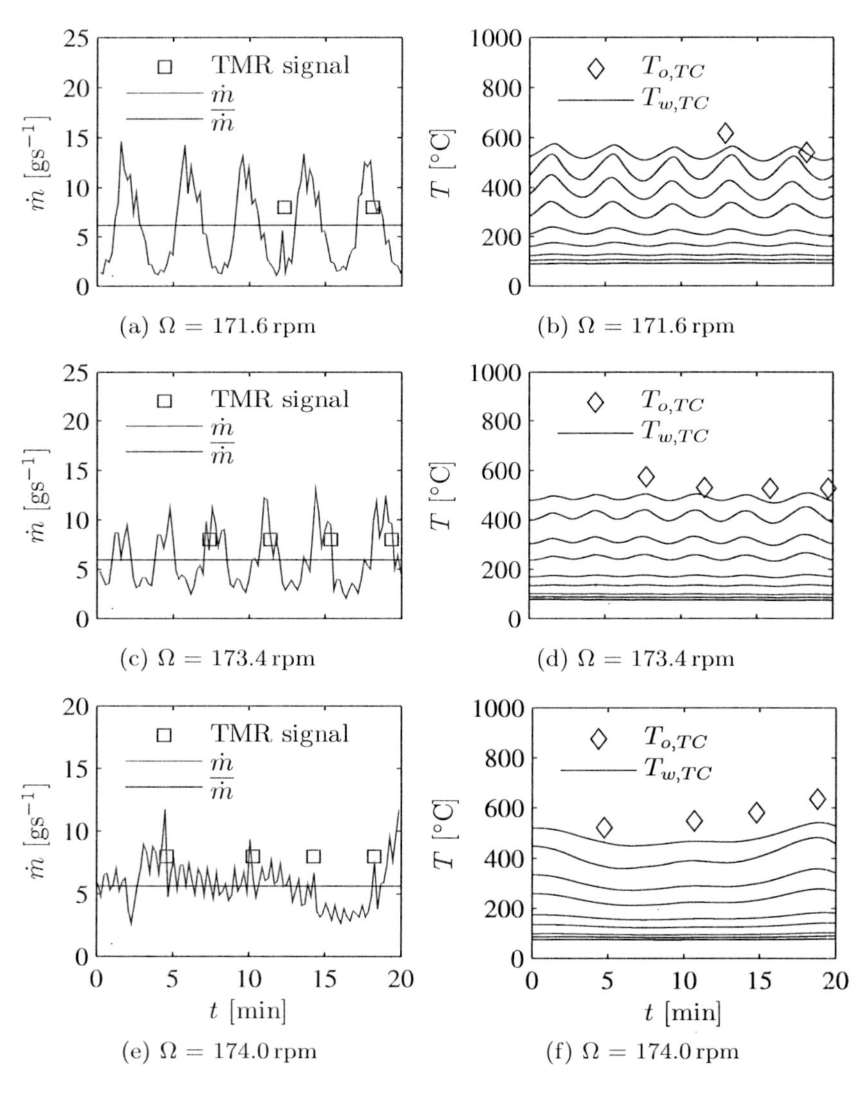

Figure 5.15: Particle temperature and mass flow rate for various Ω with insufficiently high receiver vibration, $\dot{m}_{nom} = 6\,\mathrm{g\,s^{-1}}$, $\dot{q}_{in} = 265\,\mathrm{kW\,m^{-2}}$.

lower temperatures are determined for increased mass flow rates.

Increasing the rotation speed (Figure 5.15c and 5.15d) does not diminish the oscillations, but leads to a decrease of amplitude and period which is also directly reflected in the temperature behavior. While nearly periodic oscillations occur for $\Omega = 171.6\,\mathrm{rpm}$ and $\Omega = 173.4\,\mathrm{rpm}$ irregular fluctuations prevail for the highest rotation speed of $\Omega = 174\,\mathrm{rpm}$ although the amplitude is significantly decreased (Figure 5.15e and 5.15f). Deviations in temperatures up to more than 100 °C in the outlet temperature are also present underlining the unsteady and discontinuous particle flow behavior.

5.2.2 Receiver inclination 90°

The next interesting receiver inclination being investigated is for the face-down case, where $\alpha = 90°$. When neglecting any wind influences high thermal efficiencies are expected due to minimized convection losses. Since experimental time was limited, only one mass flow rate could be properly tested. The focus lay on reaching the target outlet temperature of 900 °C at the highest possible input power.

Figure 5.16 displays the interpolated and measured wall temperatures for a mass flow rate of $\dot{m}_{nom} = 8\,\mathrm{g\,s^{-1}}$ and two input heat fluxes, $\dot{q}_{in} = 670\,\mathrm{kW\,m^{-2}}$ and $\dot{q}_{in} = 520\,\mathrm{kW\,m^{-2}}$. Interpolated temperatures in the upper figures indicate the temperature distributions to be not as homogeneous as for $\alpha = 45°$, but still relatively uniform, when considering the SD related to the mean shown by the diagrams on the right. The observation from Chapter 5.1.2 where the particle movement is presumed to be more sensitive to unfavourable influences like receiver eccentricity, unbalances and external vibrations for $\alpha = 90°$ as for $\alpha = 45°$ can be considered as verified.

Looking at the measured temperature profiles in Figure 5.16c and Figure 5.16d, it appears that the overall temperature level is

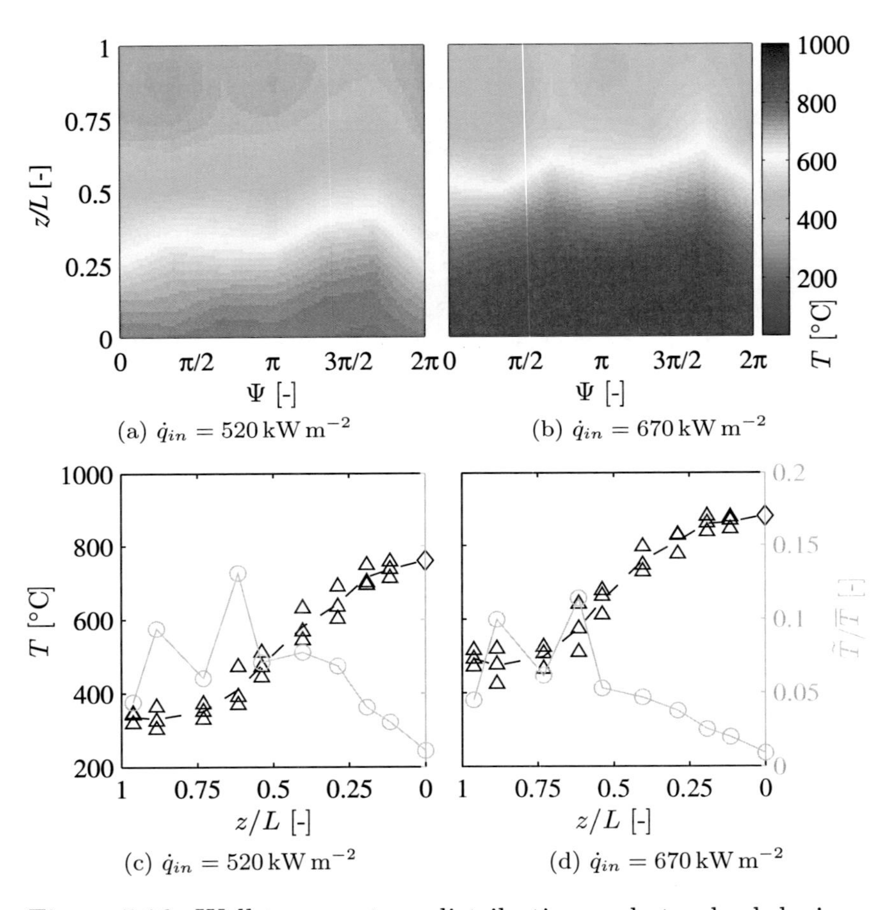

Figure 5.16: Wall temperature distribution and standard deviation for two input heat fluxes with $\dot{m}_{nom} = 8\,\mathrm{g\,s^{-1}}$. ($\triangle$) $T_{w,TC}$, ($\diamond$) $T_{o,TC}$, (-⊖-) $\tilde{T}/\overline{T}$

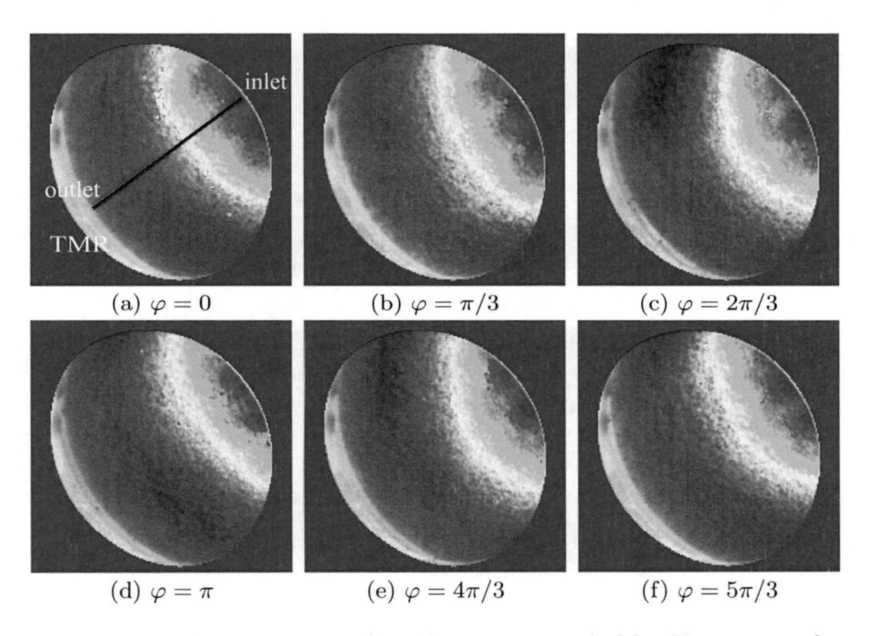

(a) $\varphi = 0$ (b) $\varphi = \pi/3$ (c) $\varphi = 2\pi/3$

(d) $\varphi = \pi$ (e) $\varphi = 4\pi/3$ (f) $\varphi = 5\pi/3$

Figure 5.17: Temperature distribution recorded by IR camera for $\dot{m}_{nom} = 8\,\mathrm{g\,s^{-1}}$ and $\dot{q}_{in} = 670\,\mathrm{kW\,m^{-2}}$. Qualitative comparison of circumferential distribution, therefore no temperature scale is shown.

in general higher compared to the $\alpha = 45°$ case. The first thermocouples close to the receiver inlet measure even temperatures up to 500 °C. As already observed in the previous chapter particles seem to be preheated before entering the receiver. However, it is surprising that considering the particle outlet temperature to be about 900 °C more as half of the temperature gradient is already reached by preheating. Caution must be taken for this interpretation though as the thermocouples do not directly measure the actual inlet temperature of the particles since they are placed behind the receiver wall. Considering the measured temperature of the feeding cone to be about 800 °C, it can be still concluded, that the particle's inlet temperature might be in the same order of magnitude of approximately 500 °C.

Qualititave comparisons of infrafred recordings to thermocouple measurements for $\dot{m}_{nom} = 8\,\mathrm{g\,s^{-1}}$ and $\dot{q}_{in} = 670\,\mathrm{kW\,m^{-2}}$ are shown in Figure 5.17. Due to space and view limitations it was not possible to observe the entire cavity wall, but only about three quarters of it. Unlike for the $\alpha = 45°$ case, the feeding cone is therefore not visible. As also indicated in the interpolation plots (Figure 5.16a and 5.16b), the overall temperature level is generally higher while the temperature distribution is quite uniform with just small deviations.

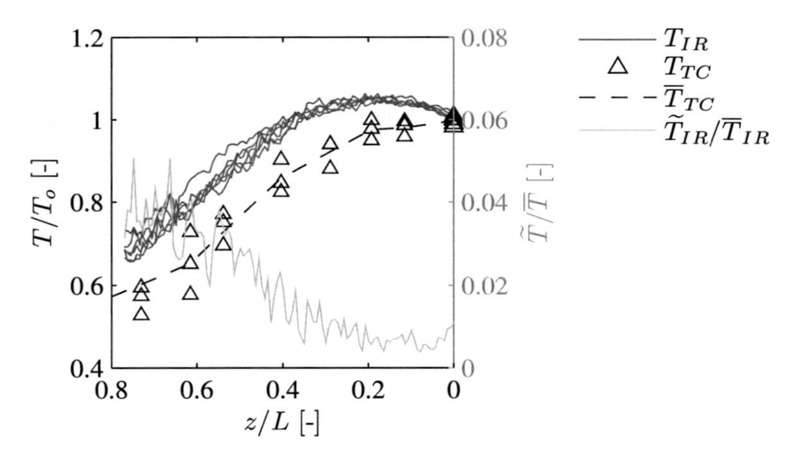

Figure 5.18: Qualitative comparison of temperature distributions along the line in Figure 5.17a to ones measured by thermocouples.

Temperature profiles along the straight line in Figure 5.17a are qualitatively compared to wall temperatures measured by thermocouples. According to Figure 5.18 temperatures measured by the IR camera are in general higher, which might indicate not negligible deviations between particle film and measured wall temperature. Moreover, considering the IR records, it appears that the particle film exhibits maximum temperatures around $z/L = 0.2$, which significantly exceed the outlet temperature. For $z/L < 0.2$ the IR temperature distribution decreases while the measured temperatures slightly increase.

5.2.3 Discussion

The presented high flux experiments have demonstrated the successfull heating of particles to the target outlet temperature of 900 °C. Measured temperature profiles of the receiver wall together with measured mass flow rates indicate a homogeneously distributed and steadily moving particle film. Qualitative comparisons with recordings of an infrared camera revealed good agreement regarding wall and film temperature distribution.

Considering the case at $\alpha = 90°$ and 900 °C outlet temperature, a receiver efficiency of about $\eta_{th} = 75\,\%$ is determined. However, this value does not represent the efficiency at design power. The CentRec prototype has been developed such that for an input power of 15 kW a flux density of 1 MW m^{-2} is achieved on the aperture. But due to aging of the lamps and positioning as well as focusing uncertainties, only part load conditions could be tested. Maximum input heat flow rates of only $\dot{Q}_{in} = 5.5$ kW [$\dot{q}_{in} = 370$ kW m^{-2}] for $\alpha = 45°$ and $\dot{Q}_{in} = 10$ kW [$\dot{q}_{in} = 670$ kW m^{-2}] for $\alpha = 90°$ could be therefore accomplished. Thermal efficiencies for the full load range are predicted by the numerical receiver model instead. Results will be presented in the subsequent chapter.

The thermal receiver performance was investigated for two relevant inclination angles, whereas the main focus was on experiments for $\alpha = 45°$. The remarkable increase of the wall temperature right after the cavity inlet is observed for both examined angles. A maximum temperature of about 500 °C has been even measured for $\alpha = 90°$ and $\dot{q}_{in} = 670$ kW m^{-2}. Despite the uncertainty of this measured temperature to be the actual particle temperature at this position, a significant preheating of the particles can be assumed. Several reasons are imaginable, such as preheating by the feeding cone (which exhibits temperatures of up to 800 °C) or by the accumulated hot air in the stagnation zone. However, the actual source could not be finally identified. For future work this observation should be thus investigated in more detail. Especially for validation purposes the accurate determination of the particle inlet temperature into the cavity is of crucial importance.

Investigations with various rotation rates at a constant mass flow rate have revealed no dependence of the particle outlet temperature on the film thickness. It is assumed that because of the small scale of the prototype, deviations in the film thickness are rather marginal leading to insignificant alterations in the outlet temperature. However, for subsequent research, the influence in larger scale receivers should be addressed as thicker particle films might be existent.

Chapter 6

Simulation

Simulations based on the model, introduced in Chapter 4, are conducted in order to get more insight in the thermal behavior of the receiver. On the one hand, the particle outlet temperature and the thermal receiver efficiency are evaluated for different operation parameters. On the other hand, the quantities of single heat transfer mechanisms are investigated in detail. A summary of all simulation parameters is given in Table 6.1.

For validation purposes, simulation results are compared to experimental measurements and are presented in the following sections. A sensitivity analysis of the model and a closing discussion about the present findings complete this chapter.

6.1 Temperature profiles

For validation purposes simulated temperature profiles of the particle layer from receiver inlet to outlet are compared to measured profiles already presented in the previous chapter. Model boundary and initial conditions, such as input power, mass flow rate and particle inlet temperature are chosen according to the corresponding experiments. Note, the particle surface temperature

Table 6.1: Input parameters for simulation according to experimental cases.

Case	α [°]	$\dot{q}_{in}$ [kW/m^2]	$\dot{Q}_{in}$ [W]	$\dot{m}$ [g/s]	T_{in} [°C]
1	45	265	3962 ± 198	9.51 ± 0.12	20
2				6.02 ± 0.20	25
3				3.94 ± 0.12	17
4				2.82 ± 0.17	19
5	45	370	5541 ± 277	9.58 ± 0.12	17
6				6.07 ± 0.16	21
7				4.16 ± 0.10	20
8				2.93 ± 0.17	24
9	90	520	7761 ± 388	7.30 ± 0.21	43
10		670	9992 ± 500	7.89 ± 0.18	36

along the receiver wall is evaluated in the simulation while in fact the receiver back wall temperature is measured in the experiments due to reasons given in Chapter 3.2.1. As no significant differences between actual particle and receiver wall temperature are expected, both quantities are taken for comparison. Moreover, the most important quantity, the particle outlet temperature can be directly compared as it is measured by the TMR.

The simulated mean flux distributions on the receiver wall $\dot{q}_{in,w}$ for the different input power levels are shown in Figure 6.1. As expected, $\dot{q}_{in,w}$ is significantly higher for cases, where $\alpha = 90°$ (red and green line) as the incoming radiation axis is nearly coaxial with the receiver axis resulting in intensified radiation of the receiver inlet region.

Figure 6.2 presents temperature profiles for various mass flow rates at $\alpha = 45°$ and $\dot{q}_{in} = 265\,\mathrm{kW\,m^{-2}}$. The mean temperature of the modelled particle lines is denoted by $\overline{T}_p$ whereas $\overline{T}_{w,exp}$ and $\overline{T}_{o,exp}$ correspond to the mean of the experimentally mea-

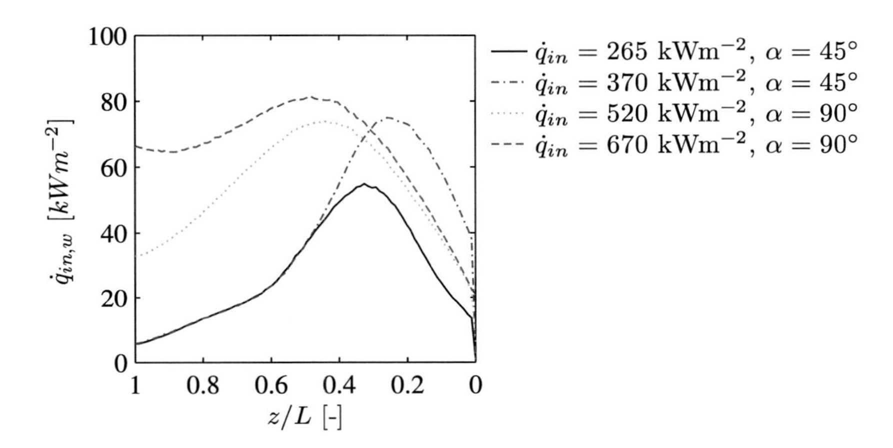

Figure 6.1: Simulated mean flux distributions on the receiver wall.

sured back wall and particle outlet temperatures, respectively. Error bars of the temperature measurement are neglected as they exhibit nearly the same size as the symbols in the diagrams. However, the uncertainties of $\dot{m}$ and $\dot{Q}_{in}$, which serve as input parameters for the simulation, are considered by an error band, denoted by the grey lines $\overline{T}_{p,err}$.

Despite the good agreement of calculated and measured outlet temperatures, the temperature distribution along the receiver is rather overestimated by simulation. For higher mass flow rates (Figure 6.2a and 6.2b), where the particle heating mostly occurs around the lower quarter of the receiver ($z/L < 0.25$), the simulated temperature exhibits a more steep increase. According to the model results, the particles are heated up to a peak around $z/L = 0.2$ from where their temperature becomes stagnant or even decreases again.

This discrepancy between the temperature profiles of simulation and experiments gets more pronounced with decreasing mass

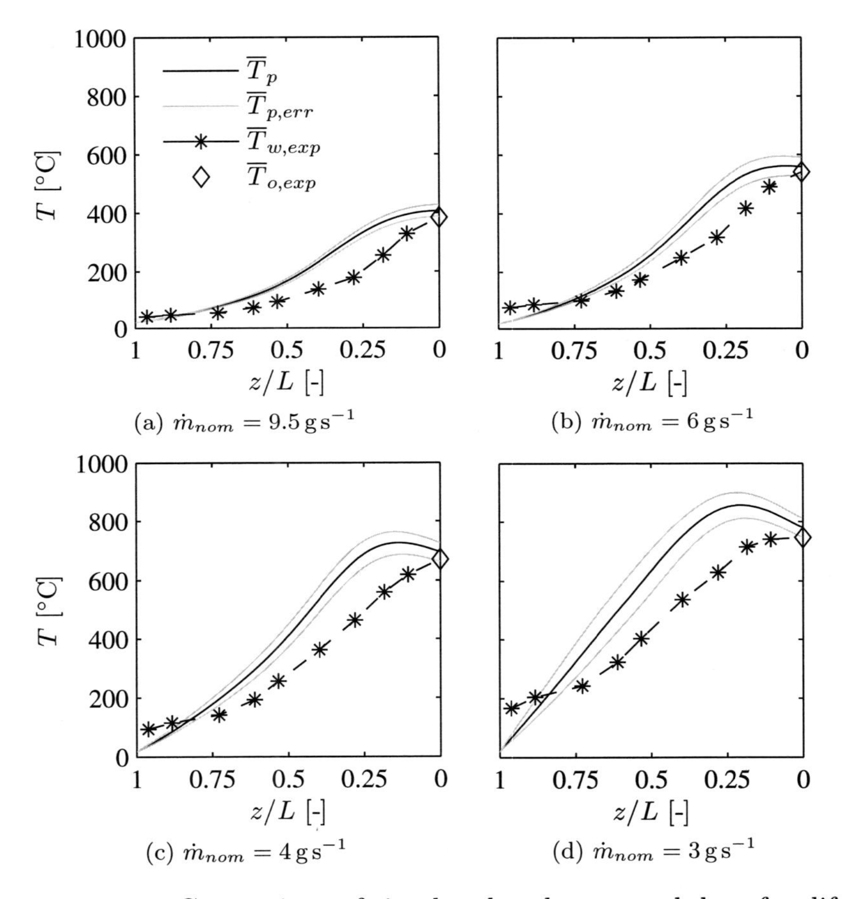

Figure 6.2: Comparison of simulated and measured data for different mass flow rates at $\dot{q}_{in} = 265\,\mathrm{kW\,m^{-2}}$ and $\alpha = 45°$. The legend in (a) is valid for all figures.

flow rate and higher overall temperatures (Figure 6.2c and 6.2d). The calculated temperature rapidly increases to a maximum that is even significantly higher as the actual particle outlet temperature. From this point it decreases again and meets $\overline{T}_{o,exp}$ quite exactly. The highest deviation between measured and simulated temperature is about 200 °C for the case with $\dot{m}_{nom} = 3\,\mathrm{g\,s^{-1}}$.

It is further observed, that the temperatures at the receiver inlet deviates increasingly with lower mass flow rates. As mentioned before the measured higher temperatures at this position could indicate a preheating of the particles that occurs somewhere outside the cavity, presumably in the feeding cone. In the simulation however, this first temperature is an initial condition for the particle lines given by the user according to the descriptions in Chapter 4.3.1. Temperatures at $z/L = 1$ can thus not be directly compared to each other.

The comparison of temperature profiles for a higher incoming heat flux of $\dot{q}_{in} = 370\,\mathrm{kW\,m^{-2}}$ are depicted in Figure 6.3. A similar tendency is clearly visible, where simulation predicts higher temperatures. Both distributions, the modelled as well as the measured one, seem not to alter with varying input power as they closely resemble the ones in Figure 6.2. However, higher temperatures are in general reached for higher input power levels.

While for the previous experiments with $\dot{q}_{in} = 265\,\mathrm{kW\,m^{-2}}$ the particle outlet temperature is still predicted well by simulation, it is slightly overestimated for $\dot{q}_{in} = 370\,\mathrm{kW\,m^{-2}}$. This is especially evident in Figure 6.3d although the deviation still lies in the expected error range. A discussion of the deviations between simulation and experiments based on a sensitivity analysis and an estimation of possible effects is given in Chapter 6.3.

Finally, measured and modelled temperature profiles $\alpha = 90°$ and $\dot{m}_{nom} = 8\,\mathrm{g\,s^{-1}}$ are compared in Figure 6.4. Next to the recurring overestimation of temperatures by simulation a significantly high discrepancy at the receiver inlet is obvious. Despite

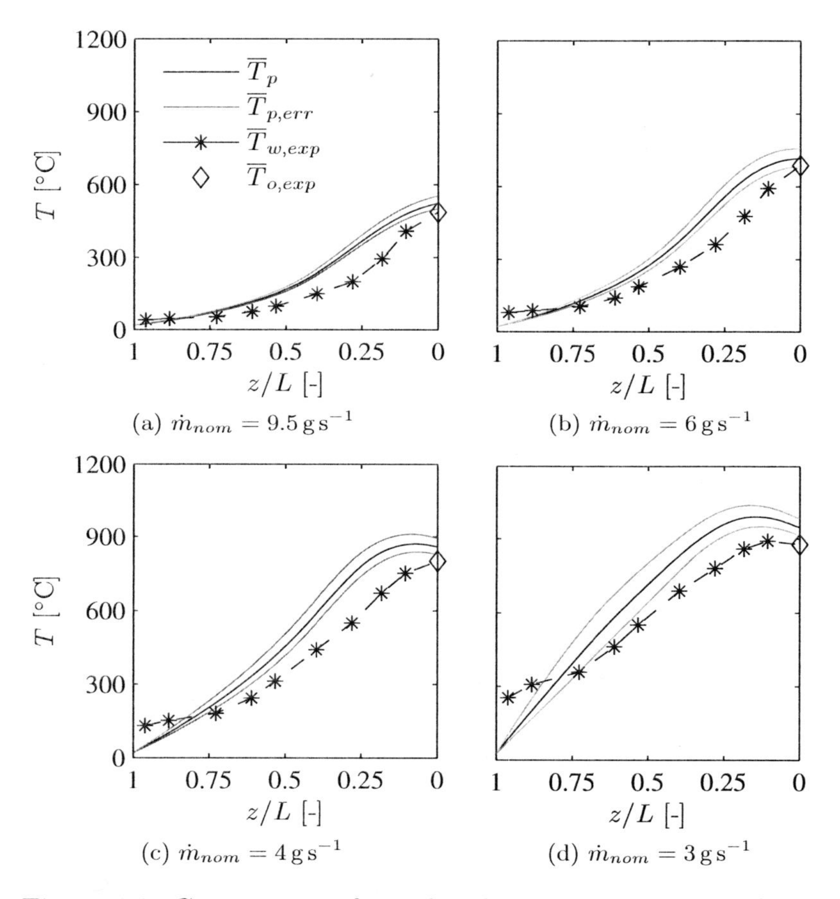

Figure 6.3: Comparison of simulated and measured data for different mass flow rates at $\dot{q}_{in} = 370\,\mathrm{kW\,m^{-2}}$ and $\alpha = 45°$. The legend in (a) is valid for all figures.

this deviation, which is due to the enormous preheating of particles in the experiment and not considered in the simulation, the particle outlet temperatures agree quite well again. Considering the temperature profiles for $z/L < 0.7$ the deviation between model and experiment tends to be decreased compared to the cases with $\alpha = 45°$. The shape of both distributions, $\overline{T}_p$ and $\overline{T}_{w,exp}$, seems to assimilate and a maximum temperature difference of $< 100\,°\mathrm{C}$ is observed in Figure 6.4b.

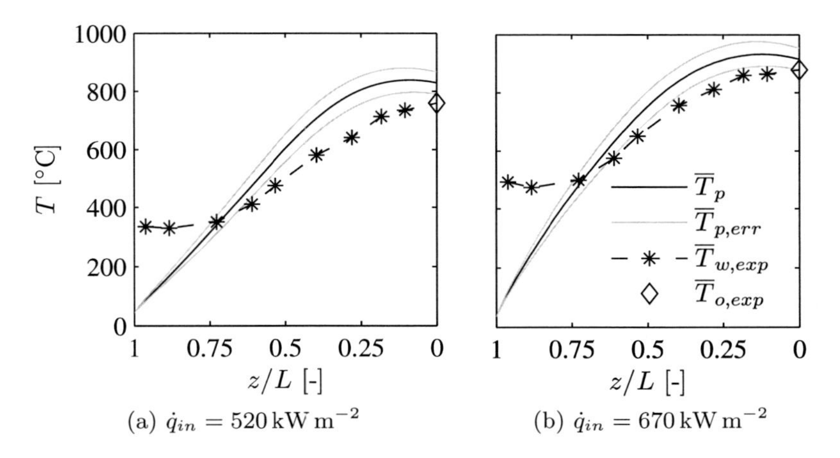

(a) $\dot{q}_{in} = 520\,\mathrm{kW\,m^{-2}}$ (b) $\dot{q}_{in} = 670\,\mathrm{kW\,m^{-2}}$

Figure 6.4: Comparison of simulated and measured data for various input heat fluxes at $\dot{m}_{nom} = 8\,\mathrm{g\,s^{-1}}$ and $\alpha = 90°$. The legend in (b) is valid for both figures.

Qualitative comparisons of measured ($\overline{T}_{w,exp}$) and simulated ($\overline{T}_p$) temperature profiles with ones recorded by the IR camera ($\overline{T}_{IR}$) are exemplarily shown for case 8 and 10 in Figure 6.5. As the IR camera is installed such, that it directly looks through the open aperture onto the cavity walls, it should be able to measure the actual film temperature. IR data should be hence more suitable for comparison to simulated data. However, due to several uncertainty factors, like the particle's emissivity is not exactly known and unidentified reflection errors, only qualitative comparisons are possible. The temperatures of each method are therefore non-dimensionalized by their corresponding outlet temperature.

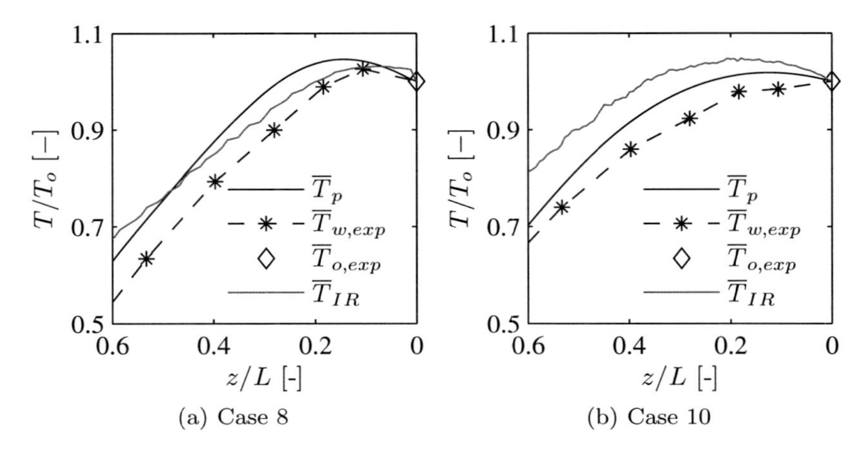

(a) Case 8

(b) Case 10

Figure 6.5: Qualitative comparison of simulated and measured temperature distribution to IR recordings for case 8 and 10.

Examing Figure 6.5a for case 8, $\overline{T}_{IR}$ exhibits in general a higher distribution than $\overline{T}_{w,exp}$, but it is still lower than the predicted temperature distribution by the simulation ($\overline{T}_p$) for 0.1 $< z/L <$ 0.5. On the contrary, $\overline{T}_{IR}$ is significantly higher compared to simulated and measured distributions for case 10 as seen in Figure 6.5b. Both figures indicate that the real particle film temperature might be indeed higher than the ones measured by the thermocouples, resembling more the simulated temperatures. However, due to the qualitative nature of the IR camera measurements the actual film temperature cannot be finally determined. An attempt to estimate the difference between real particle film and measured wall temperature is thus given in Chapter 6.3.3.

6.2 Receiver efficiency

The main simulation results for experiments at $\alpha = 45°$ are in summary presented in Figure 6.6. On the left side, the outlet temperature with respect to particle mass flow rate is separately shown again along with the absorbed heat flow rate $\dot{Q}_{abs}$, that is calculated according to (3.2). Thermal receiver efficiencies are compared on the figure's right side and the four major losses relative to the input power are exposed in detail.

Comparing simulation to experimental data (denoted by discrete symbols), good agreement within the defined error ranges can be found in general. As expected, the particle outlet temperature increases with decreasing mass flow rate for a constant input power. The overall absorbed heat however, decreases along with the receiver efficiency as thermal losses predominate with higher temperatures. This behavior is clearly indicated for both input power levels.

Examining the heat losses in more detail, it is clearly revealed that the most dominant loss is due to radiation. It increases up to over 22 % of the input power for $T_o = 900\,°C$, which is not surprising since the temperature is considered with the fourth power in the Stefan-Boltzmann law. As the receiver is inclined with $\alpha = 45°$, the second highest loss is expected to be convection with a significant part of 10 % for the highest outlet temperature. Conduction losses play a rather minor role with a maximum of about 7 % and are strongly dependent on the receiver design and set-up. It is imaginable to minimize them to only 1 % in future scale-up CentRec designs. Optical losses are observed to be independent from the particle mass flow rate and temperature as they are only affected by incicent heat flux and receiver geometry. With the assumed absorptivity of $a = 0.89$ and the present cavity geometry $\dot{Q}_{opt}$ is calculated to be 2 % and 3 % for $\dot{q}_{in} = 265\,\mathrm{kW\,m^{-2}}$ and $\dot{q}_{in} = 370\,\mathrm{kW\,m^{-2}}$, respectively.

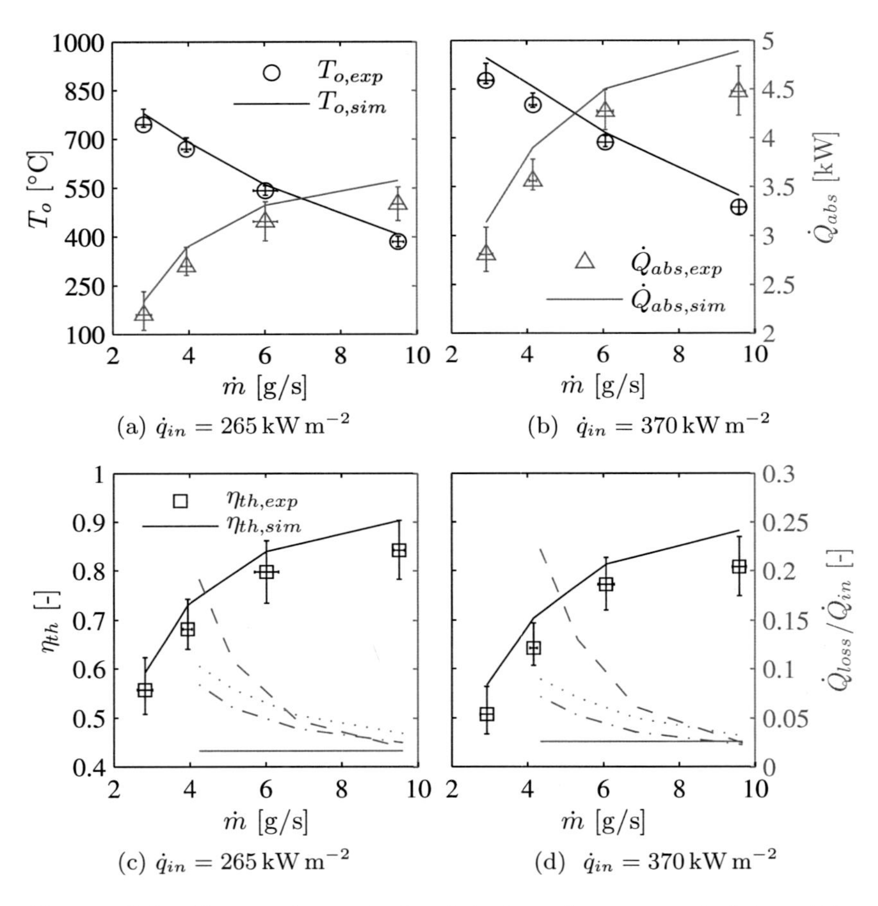

Figure 6.6: Comparison of simulated and measured data with respect to particle mass flow rate for considered input heat fluxes at $\alpha = 45°$. Relative heat losses calculated within the model are presented in the lower graphs on the right-hand y-axis: (—) optical, (- - -) radiation, (· · ·) convection, (- · -) conduction.

Summing up all heat losses, the highest thermal receiver efficiency is reached for $\dot{m} = 9.51\,\mathrm{g\,s^{-1}}$ and $\overline{T}_o = 407\,°\mathrm{C}$ with $\eta_{th} = 90\,\%$. For a particle outlet temperature of $\overline{T}_o = 900\,°\mathrm{C}$, η_{th} is calculated to be around 56 % which seems rather low. However, as the receiver has been originally designed for an incoming heat flux of $\dot{q}_{in} = 1\,\mathrm{MW\,m^{-2}}$, but just one third could be achieved due to limitations of the solar simulator, this efficiency corresponds to a part load case.

Despite the deviations in the temperature profiles between simulation and experiments, the introduced receiver model can be seen as validated as the particle outlet temperature and the integral value $\dot{Q}_{abs}$ agree quite well. An extrapolation of the thermal receiver efficiency with $T_o = 900\,°\mathrm{C}$ is therefore justified for higher load conditions. Simulations are conducted for the considered receiver inclinations $\alpha = 45°$ and $\alpha = 90°$ and presented in Figure 6.7.

It is revealed, that a receiver efficiency of about $\eta_{th} = 87\,\%$ can be achieved under full load conditions at the design heat flux of $\dot{q}_{in} = 1\,\mathrm{MW\,m^{-2}}$. It rapidly decreases with lower input power as the share of radiation losses increases. The efficiency distribution between both tilt angles appears to look quite similar although significant lower convection losses are exhibited for the $\alpha = 90°$ case in Figure 6.7b. This effect is due to different incoming heat flux distributions of the solar simulators leading to a generally higher temperature level of the receiver in $\alpha = 90°$ position. Increased radiation losses are the consequences whereas lower convection losses are compensated. In cases, where an identical heat flux distribution is applied, a face down receiver might yield higher efficiencies.

However, despite the influence of different incoming flux distributions, it seems that with increasing input power the effect of inclination angle diminishes. Convection losses become less pronounced with higher input power resulting in nearly same ef-

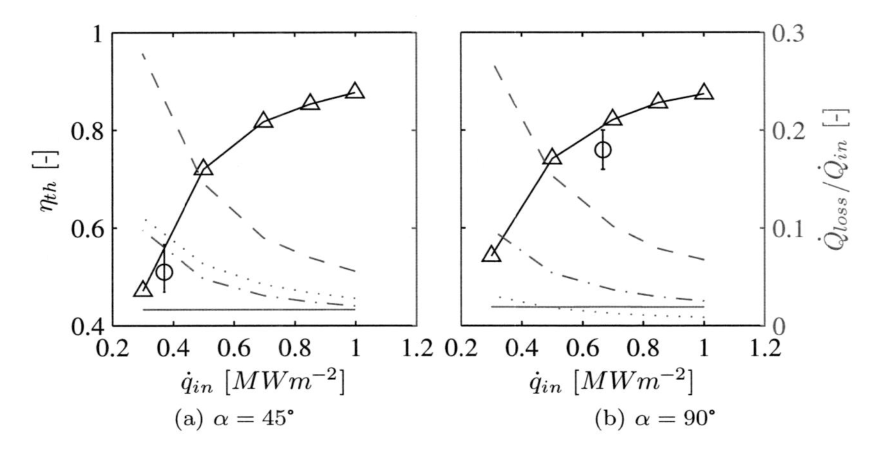

(a) $\alpha = 45°$ (b) $\alpha = 90°$

Figure 6.7: Thermal efficiency versus load conditions where $\overline{T}_o = 900\,°C$. Full load is at $\dot{q}_{in} = 1\,\mathrm{MW\,m^{-2}}$: (-△-) $\eta_{th,sim}$, (○) $\eta_{th,exp}$. Relative heat losses are presented on the right-hand y-axis: (—) optical, (- - -) radiation, (· · ·) convection and (- · -) conduction.

ficiencies for both inclination angles at full load condition. The effect of α and convection losses increases for lower part load states. In the case of 30 % part load the efficiency for $\alpha = 45°$ is decreased to about $\eta_{th} = 47\,\%$ while for $\alpha = 90°$ it is still around $\eta_{th} = 54\,\%$. Note, these observations are only valid when no wind effects on convection are considered.

6.3 Sensitivity analysis

The deviations between simulated and measured temperature distributions lead to the suspicion that there are model assumptions which might not be entirely correct, insufficiently accurate or even missing. In order to get an impression of the quantitative influence of relevant parameters on the simulation results, such as material properties, incoming heat flux distribution, conduction and convection effects, a thorough sensitivity analysis and the attempt to estimate the real particle film temperature are subsequently conducted and discussed.

6.3.1 Material properties

For accurate simulation results, the implementation of correct material properties is essential. The emissivity ε and the absorptivity a of the particles influence the calculation of two main heat losses, radiation and optical reflection, respectively. Several simulations with varied values of ε and a are thus considered, in order to get an impression of their quantitative effect on the overall model. Reference values are chosen to be $a = 0.89$ and $\varepsilon = 0.82$, defined in Chapter 4.3.2 and according to the measurements by Siegel [65].

Figure 6.8 presents the effect of ε on temperature profiles and heat losses for case 8 from Table 6.1 and a constant absorptivity of $a = 0.89$. For comparison reasons experimental data are shown additionally in Figure 6.8a. A nearly negligible influence of ε on the particle temperature is clearly identified as variations between the temperature distributions are quite small.

A more detailed look into the alteration of particle outlet temperature and heat losses is given in Figure 6.8b. As can be seen, increasing ε leads to lower outlet temperatures since thermal heat losses are increased. Considering $\varepsilon = 0.82$ to be the reference value the relative changes of the heat losses for different ε is generally quite low. A maximum increase of about 4 % for radiation losses is indicated for $\varepsilon = 0.97$. Optical losses are not affected at all as they are only dependend on the absorptivity coefficient.

Similar conclusions can be derived when looking at Figure 6.9a which exhibits the dependence of the temperature distribution for different a and a constant emissivity of $\varepsilon = 0.82$. Only small deviations between the temperature profiles around

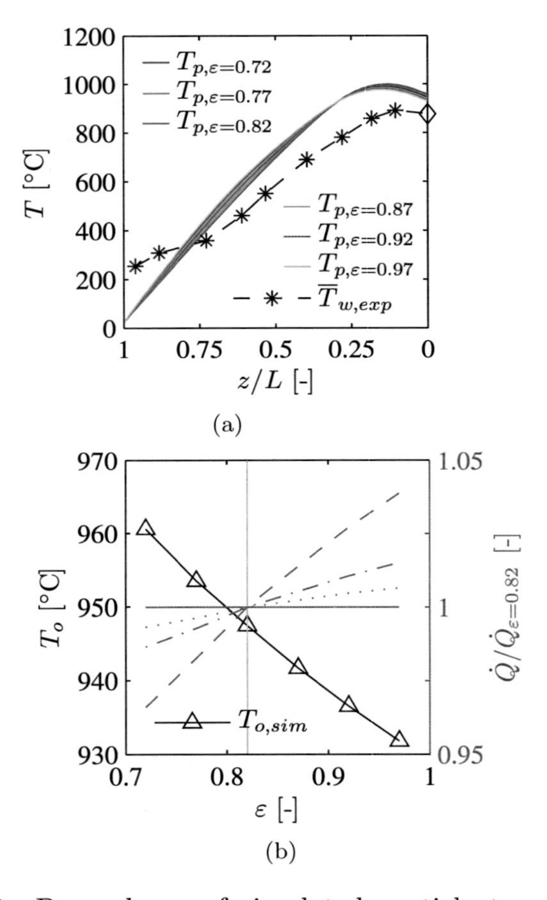

Figure 6.8: Dependence of simulated particle temperature on the emissivity ε for constant $a = 0.89$. Relative heat losses are presented in (b) on the right-hand y-axis: (—) optical, (- - -) radiation, (· · ·) convection, (- · -) conduction. $\dot{m}_{nom} = 3\,\mathrm{g\,s^{-1}}$, $\alpha = 45°$, $\dot{q}_{in} = 370\,\mathrm{kW\,m^{-2}}$.

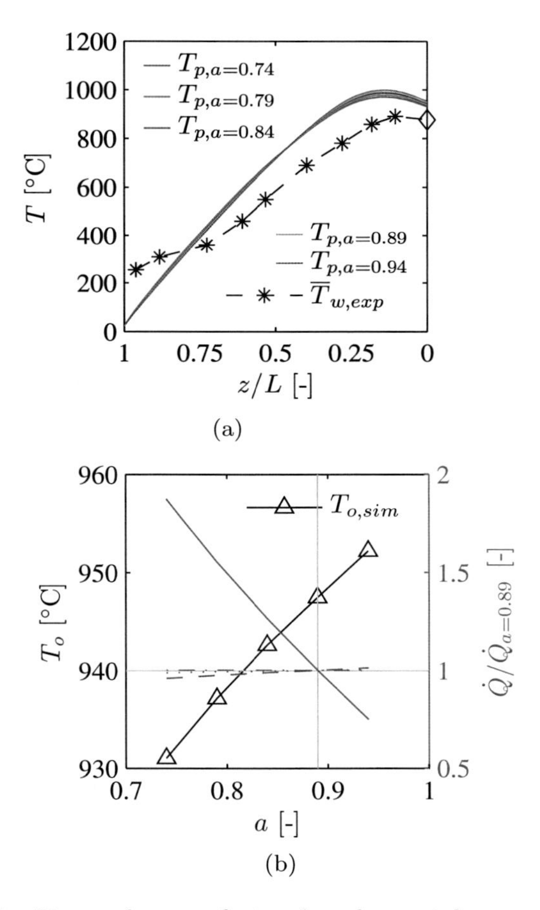

Figure 6.9: Dependence of simulated particle temperature on the absorptivity a for constant $\varepsilon = 0.82$. Relative heat losses are presented in (b) on the right-hand y-axis: (—) optical, (- - -) radiation, (· · ·) convection, (- · -) conduction. $\dot{m}_{nom} = 3\,\mathrm{g\,s^{-1}}$, $\alpha = 45°$, $\dot{q}_{in} = 370\,\mathrm{kW\,m^{-2}}$.

$z/L = 0.2$ are visible. According to Figure 6.9b the outlet temperature decreases with lower a, which is quite understood as the optical losses are highly increased. However, despite its relative increase of over 85 % for $a = 0.74$ the influence of a on the general simulation results is relative small as optical losses play a very minor role among the overall receiver losses (see Figure 6.6d). All other thermal losses are therefore nearly not affected by a. Radiation losses are slightly decreased due to the lower temperatures.

6.3.2 Heat flux distribution

The influence of the incoming heat flux distribution on the simulation results is investigated. As described in Chapter 4.5 it is quite complex to accurately model the input power coming from the solar simulator. Although the comparison of modeled and measured flux distributions yield good agreement, uncertainties might be still present as these comparisons are more qualitative. Moreover, as the measured 2D flux map does not give any information about the ray distribution, identical flux maps does not mean identical flux distributions inside the receiver.

Looking at Figure 6.3a and 6.3b for example, which indicate the particle heating to mostly occur around the lower quarter of the receiver, it can be assumed that the irradiation is more concentrated in the outlet region as well. Based on these considerations a new arbitrary heat flux distribution is modeled with its highest peak close to the receiver outlet. This is done by shifting the receiver 80 mm backwards along the z-axis and adjusting the heat flux factor to end up with the same input power in the aperture plane. Both heat flux distributions, the original ($\dot{q}_{in,orig}$) and the varied one ($\dot{q}_{in,var}$), are plotted in Figure 6.10.

The distribution of $\dot{q}_{in,var}$ is visibly lower as the one of $\dot{q}_{in,orig}$ for $z/L > 0.2$. From the lower third of the receiver, $z/L = 0.3$, a steep increase occurs in the varied distribution up to a narrow peak close to the receiver outlet, where $\dot{q}_{in,var} = 12 \times 10^4\,\mathrm{W\,m^{-2}}$. In contrast, the maximum of $\dot{q}_{in,orig}$ is significantly lower and already at $z/L \approx 0.3$ and decreases again at the lower third of the receiver. Note, $\dot{q}_{in,var}$ is arbitrarily chosen in order to demonstrate the sensitivity of the model to the incoming flux distribution. There is no reasonable argument why $\dot{q}_{in,var}$ should exactly look like this.

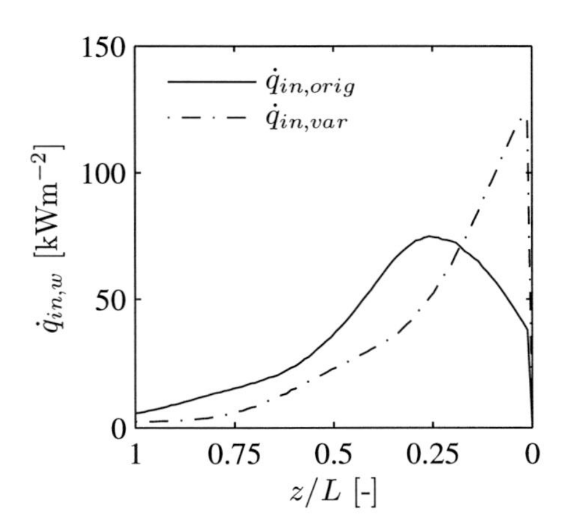

Figure 6.10: Altered heat flux distribution. $\dot{q}_{in} = 370\,\mathrm{kW\,m^{-2}}$, $\alpha = 45°$.

Simulations are conducted where $\dot{q}_{in,var}$ is applied exemplarily to case 5 and 8 from Table 6.1. The corresponding results of temperature profiles and receiver losses are shown in Figure 6.11. The effect of $\dot{q}_{in,var}$ is quite significant. Considering case 5 the varied heat flux distribution leads to a simulated temperature distribution that almost perfectly matches the measured one (Figure 6.11a). Examining the altered receiver losses in Figure 6.11b an extensive increase of the relative optical losses from 2.5 % for the original to 7.5 % for the varied heat flux distribution is apparent. As thermal losses remain nearly unchanged and all on

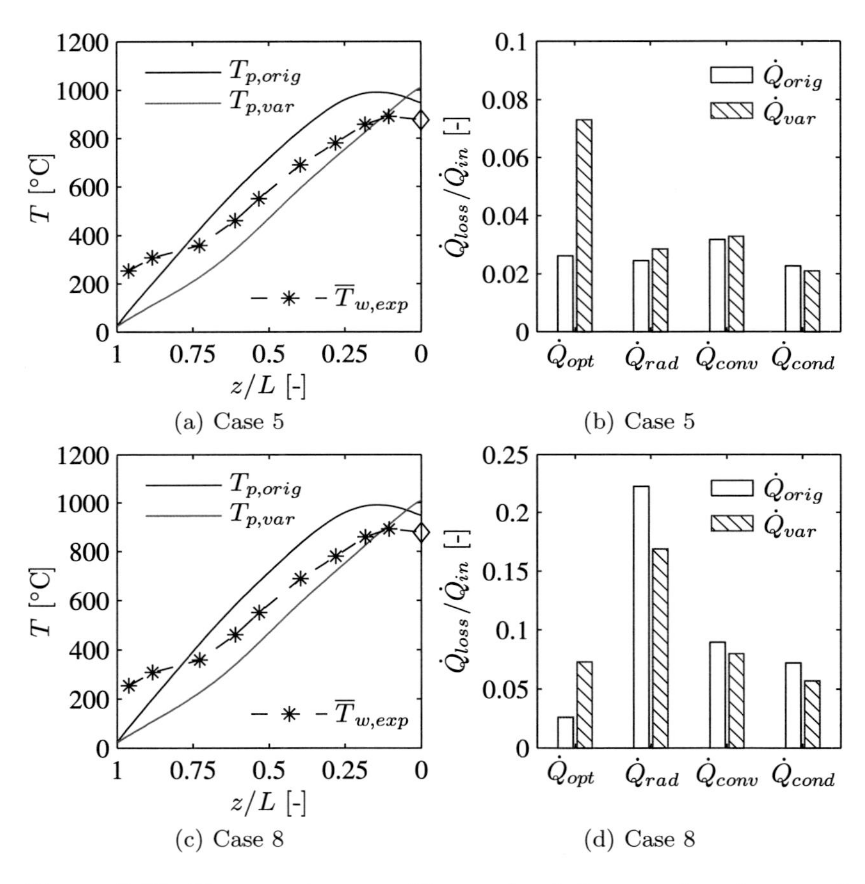

Figure 6.11: Temperature profiles and relative receiver losses of case 5 and 8 for two different heat flux distributions.

the same level of 2 to 3 % due to the low temperatures, $\dot{Q}_{opt,var}$ is now the highest share in the overall receiver losses.

For case 8 however, the simulated temperature distribution for $\dot{q}_{in,var}$ is now quite different to the measured one indicated in Figure 6.11c. The particle outlet temperature is calculated to be more as 100 °C higher. This observation can be explained by investigating the receiver losses in Figure 6.11d. As the temperature level of the particles increases thermal losses, especially radiation losses, predominate over the optical losses. Due to the defined particle inlet temperature and the distribution of $\dot{q}_{in,var}$ the temperature profile $T_{p,var}$ is generally lower compared to the original simulated and measured one. Thermal losses are therefore decreased leading to more absorbed heat in the particles and accordingly higher outlet temperatures.

Apparently, the incoming flux distribution essentially affects the temperature profile inside the receiver. However, the discrepancies between simulation and experiment can not be fully explained by an uncertain flux distribution. The preheating of the particles seems to play an important role which is not considered in the model so far. Looking at Figure 6.11c the simulated temperature level would be overall higher if the particle inlet temperature would be adjusted to the temperature measured by the first thermocouple at the receiver inlet. Higher temperatures lead to higher thermal losses and therefore reduced outlet temperatures which would fit more to the experimentally measured ones. The heat transfer mechanisms causing the particle preheating could not be experimentally determined with the present set-up. In the subsequent sections two imaginable possibilities are thus investigated and their quantitative effect roughly estimated.

6.3.3 Conduction effects

The increased temperatures right at the cavity inlet which lead to the assumption of preheated particles, can be caused by several

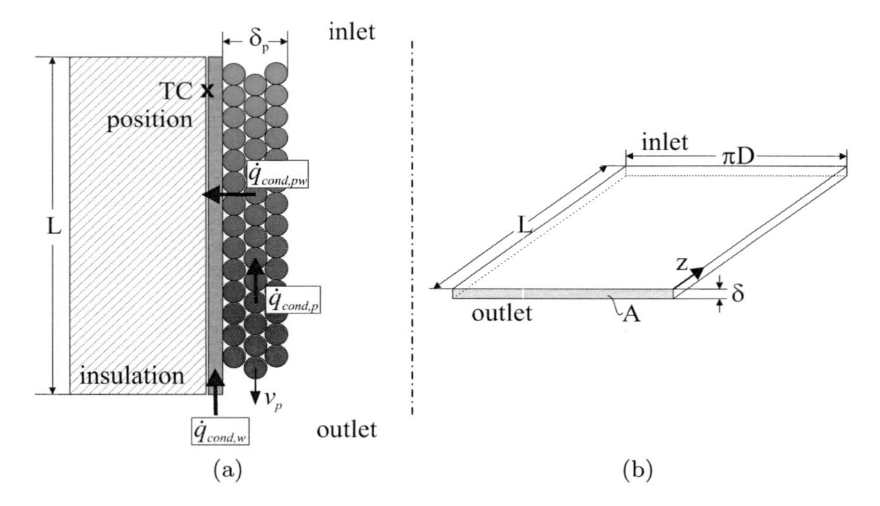

Figure 6.12: Sketch of possible conduction effects which might happen during experiments, but are not considered in the simulation.

effects. In this section, the order of magnitude of two possible conduction effects are estimated. According to Figure 6.12a, there is on the one hand the conduction through the particle layer itself contrary to there moving direction, $\dot{Q}_{cond,p}$. On the other hand, conduction through the receiver wall from the hot part at the outlet to the cold part at the inlet, $\dot{Q}_{cond,w}$, is considered.

For estimation purposes the particle layer and receiver wall are unwound as rectangles as pictured in Figure 6.12b. The conducted heat through area A along the z-axis is calculated as one-dimensional, steady-state conduction with

$$\dot{Q}_{cond,z} = \frac{\lambda}{L} \pi D \delta (T_o - T_{in}). \tag{6.1}$$

Considering the case with the highest temperature level and assume $T_o = 900\,°\mathrm{C}$ and $T_{in} = 20\,°\mathrm{C}$. The effective conductivity of the particle layer is taken according to measurements by Baumann et al. [9] for a mean temperature of $\overline{T} = 460\,°\mathrm{C}$, $\lambda_p = 0.54\,\mathrm{W\,m^{-1}\,K^{-1}}$. Supposing a particle layer thickness of 2 to 3 particles, $\delta_p = 4\,\mathrm{mm}$, $\dot{Q}_{cond,p}$ is calculated to be less than 4 W. Conduction through the particle layer is therefore negligible. Even assuming thicker particle layers would not change much, as the thickness δ is considered linearly in (6.1).

The same approach is used for estimation of $\dot{Q}_{cond,w}$. The cavity is made of a high-temperature alloy Inconel 617 with $\lambda_w = 15.8\,\mathrm{W\,m^{-1}\,K^{-1}}$ [5]. With a wall thickness of $\delta_w = 2\,\mathrm{mm}$, the conducted heat through the cavity wall is evaluated to be about 58 W, which is less than 1 % of the overall input power and can be therefore also neglected.

The remaining conductive effect, sketched in Figure 6.12, is less investigated with regard to the preheating of particles but rather to the measurement method of the receiver wall temperature. As the particle film layer should not be disturbed, thermocouples are placed behind the cavity, on the inner insulation wall. It was assumed that the temperatures measured at these positions do not differ significantly from the temperature in the particle film. However, considering the comparison of simulated and measured temperature profiles, a more detailed look into this presumption should be taken. A rough estimation of the real particle temperature is conducted in the following.

Preliminary experiments without particle film have revealed that the temperature of the inner cavity wall is constantly about 10 K higher as measured by the thermocouples installed on the inner insulation wall. This difference is already considered as a systematic error in the uncertainty analysis. As observed in Chapter 5.1.3 the particle film layer can be several particles thick. To estimate the temperature of the topmost particle layer conduction through the film in radial direction is examined.

Since the heat transfer between particle layer and the inner cavity wall is difficult to estimate due to many unknown parameters, such as particle velocity, particle layer thickness and wall roughness, it is assumed that the inner wall temperature is nearly equal to the outermost particle layer temperature.

Considering case 8 from Table 6.1, the simulated conduction losses through the insulation are about $\dot{Q}_{cond} = 400\,\mathrm{W}$. Suppose conservatively that $\dot{Q}_{cond}$ is solely lost through the insulation block the conducted heat flux can be calculated with

$$\dot{q}_{cond,pw} = \frac{\dot{Q}_{cond}}{A_m} = \frac{\lambda_m}{\delta_p}(T_p - T_w). \tag{6.2}$$

A_m denotes the mean of the inner insulation wall A_{ii} and the outer receiver wall area A_a and is determined by the conducting area of a hollow cylinder, $A_m = (A_a - A_{ii})/\ln(A_a/A_{ii})$. The resulting temperature difference between the inner cavity wall and the topmost particle layer depending on the film thickness δ_p and the temperature dependent particle conductivity λ is presented in Figure 6.13 for $\dot{q}_{cond,pw} = 1986\,\mathrm{W\,m^{-2}}$. δ_p is chosen from 2 mm for a mininum layer of 1 to 2 particles and a maximum layer of about 14 mm which is the opening width of the feeding cone. Thicker particle layers are not possible as the feeding cone would be blocked then.

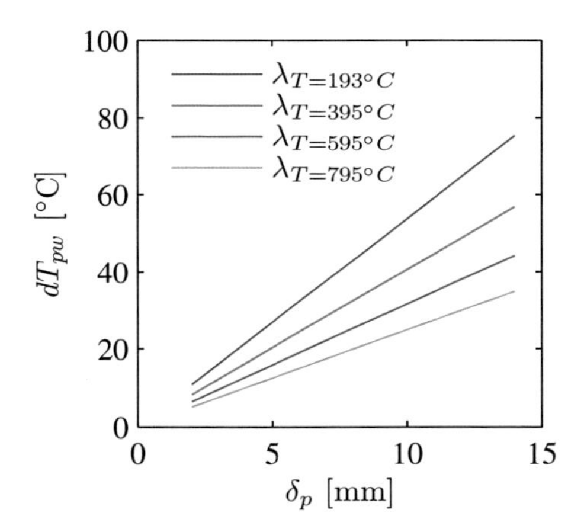

Figure 6.13: Temperature difference between cavity inner wall and topmost particle layer with respect to particle layer thickness δ_p for various temperature dependent conductivity λ and a conducted heat flux of $\dot{q}_{cond,pw} = 1986\,\mathrm{W\,m^{-2}}$.

With increasing film thickness, the temperature difference between the layers dT_{pw} increases. Decreased temperature levels lead to lower λ and therefore increased dT_{pw} as well. A maximum temperature deviation of about 70 °C for maximum δ_p and mininum λ is determined. The calculations however are conducted very conservatively. Compare to observations from Chapter 5.1.2 it is rather unlikely that the particle layer will reach its maximum over the whole receiver length. A film thickness of 3 to 4 particles with $\delta_p \approx 5\,\mathrm{mm}$ would be a more realistic case. Moreover, the conduction heat flux $\dot{q}_{cond,pw}$ is supposedly assumed too high. Due to the very low conductivity of the insulation material [2], the overall conductive losses of the receiver occur probably mostly through thermal bridges from the holding rings and mounting gaps between the single insulation blocks.

It can be concluded that the temperature difference between actual particle temperature of the topmost layer and the measured one by the thermocouples is roughly estimated about 20 to 30 °C, which is rather low and does not explain deviations of 100 to 200 °C. However, as discussed above, there are still a lot of uncertainties which could not be clarified yet. For validation purposes the present methods regarding the accurate measurement of particle surface temperatures should be improved in future work.

6.3.4 Convection effects

As it is demonstrated in Chapter 6.3.3, the considered conduction effects seem to be of minor importance. Another thinkable effect is the preheating of particles by hot air which is accumulated in the stagnation zone from Clausing's theory [18]. Since the present receiver model does not consider any CFD calculations, the model is extended as follows.

First, the required heat flow rate for the particle preheating is roughly estimated with $\dot{Q}_{prh} = \dot{m}c_p(T_{fc} - T_{in})$ with T_{fc} as the measured temperature of the feeding cone. According to $h_{prh} = \dot{Q}_{prh}/(A_{cav}(\overline{T}_w - T_\infty))$ an imaginary heat transfer coefficient is calculated then and applied as an additional boundary condition to the inner wall surface of the cavity. Through h_{prh} heat is subtracted from the particle elements in the model and redistributed by applying a higher particle inlet temperature T_{in} which is chosen according to the temperature measured by the first thermocouple close to the receiver inlet.

Simulated temperature profiles of this extended model compared to profiles of experiments and the original model are shown in Figure 6.14 for case 8 and 10. Significant deviations are observed. Compared to the experimental measurements a much stronger temperature increase is exhibited in the simulation resulting in overall calculated higher temperatures. Due to the higher T_{in} the temperature distribution is generally overestimated by the extended model. The substracted heat by the additional heat transfer coefficient appears to not affect much local temperatures but more the overall energy balance as in both cases the particle outlet temperatures agree quite well again.

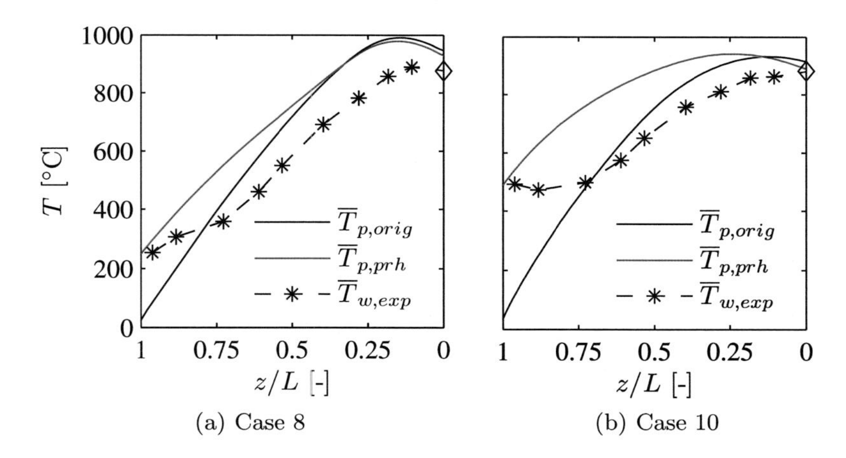

Figure 6.14: Simulated temperature profiles for case 8 and 10, when particle preheating is considered.

It seems, that the overall heat losses of the receiver are estimated properly by the simulation, but the quantity of the single loss mechanism might be not modelled correctly. Comparing measured and calculated temperature profiles radiation losses are propably overestimated by simulation. This is quite simple to prove as the radiated heat can be evaluated based on the measured wall temperatures. However, as discussed above, the wall temperatures measured by the thermocouples do not have to directly correspond to the actual particle temperatures. Conduction losses of the receiver have been experimentally measured and can be considered as known. The highest uncertainties exist for convection and optical losses. Despite the investigations in Chapter 2, where rotation effects are considered to be negligible, other factors such as the collecting ring, which is mounted right in front of the aperture, might have a strong influence. Although optical losses seem to play a minor role, their actual quantity is quite uncertain due to the complexity of accurately modelling the incoming heat flux distribution of the solar simulator.

6.4 Discussion

The validation of the numerical receiver model with experimental data has revealed good agreement in the particle outlet temperature, but significant deviations in the wall temperature profiles. Several possible reasons are extensively examined and evaluated in the sensitivity analysis. The strongest influence on the results is exerted by the incoming heat flux distribution as its modelling is afflicted with high uncertainties. The exact position of the receiver relative to the lamps is very difficult to determine as the lamps are adjusted individually for both examined inclination angles. Although the adjustment is considered by adapting the modelled to the measured flux maps on the target, these comparisons are more of a qualitative nature.

The preheating of the particles before they enter the cavity is estimated by conduction and convection effects. The influence of the feeding cone is not examined in detail as the heat transfer coefficient between particle and feeding cone is completely unknown. Moreover, simulations with altered feeding cone boundary conditions have indicated no significant influence on overall model results.

Due to the arrangement of the thermocouples along the cavity, the temperature of its back wall is actually measured while the particle surface temperature is evaluated in the simulation. Deviations between particle surface and wall temperature seem to be not negligible. The particle film thickness might play an important role, its actual influence however could not be finally identified. For future work and especially for model validation, methods should be thus developed and installed in order to be able to measure the actual film temperature together with the accurate determination of the particle inlet temperature.

Chapter 7

Conclusions

The Centrifugal Particle Receiver concept is considered to be a promising way for high thermal receiver efficiencies at temperatures up to 1000 °C, assuring cost reduction in solar power tower applications. It basically consists of a fast rotating cylindrical cavity which can be inclined 10 to 90° to the horizontal. Sintered bauxite particles of 1 mm diameter, used as heat transfer and storage medium at the same time, are forced against the cavity wall by centrifugal acceleration. An optically dense and thin particle film is formed, that slowly moves along the receiver axis while gradually being heated by incoming radiation. The target particle outlet temperature of 900 °C is achieved for all load conditions by setting a corresponding mass flow rate and adjusting the particle retention time by controlling the receiver rotation rate.

The main objectives of the present work are the demonstration of the general feasibility of the proposed concept, the determination and investigation of relevant operational parameters and the evaluation of the thermal receiver performance. A prototype in laboratory scale and a numerical receiver model were thus developed and extensively tested.

Initial experiments were devoted to the question of the simple receiver controllability. A high-speed camera was used in order to qualitatively investigate the particle film behaviour depending on various parameters, such as mass flow rate, rotation rate and inclination angle. The experiments revealed, that for each mass flow rate there exists one critical minimum rotation speed Ω_{crit} for which the development of a dense particle film is possible. For the considered relative mass flow rates of $\dot{m}^* \leq 0.036\,\mathrm{s}^{-1}$ minimum rotation rates of $\Omega^*_{crit} \geq 2.35$ are necessary. Moreover, Ω^*_{crit} increases with higher inclination angle as the fraction of gravitation in axial direction grows with increasing α.

The development of a dense moving particle film has been successfully demonstrated. However, due to particle accumulation effects an inhomogeneously moving film was detected during the experiments, resulting in mass flow rate measurements with periodically repeating oscillations. An external vibration via an additional vibration wheel had to be implemented in order to achieve a steadily moving particle film.

Depending on the load condition the mass flow rate and the particle retention time t_{ret} must be adjusted for constant outlet temperatures. The controllability of t_{ret} by regulating the rotation speed has been proven by evaluating the retention time for various rotation rates at a certain mass flow rate. Increasing Ω leads to increased retention times while the particle film remain dense and homogeneously moving.

Experiments in the High-Flux Solar Simulator at the DLR in Cologne were conducted to investigate the overall thermal performance of the proposed receiver concept. Thermocouples were distributed along the receiver wall in order to measure its thermal behavior. The particle outlet temperature was determined using a specially developed measurement system that was integrated in the rotating part. An extensive number of experiments were conducted varying the incoming heat flux, the particle mass flow rate and the receiver inclination. Results of exemplary tests were presented showing the successfull achievement of the target particle outlet temperature of $T_o = 900\,°\mathrm{C}$. Moreover, measured temperature profiles of the receiver wall together with measured mass flow rates indicate a homogeneously distributed and steady moving particle film. Qualitative comparisons with recordings by an infrared camera reveal good agreement regarding wall and film temperature distribution.

Due to lamp aging and imprecise receiver positioning relative to the lamps, the full load condition at $\dot{Q}_{in} = 15\,\mathrm{kW}$ (corresponding to $\dot{q}_{in} = 1\,\mathrm{MW\,m^{-2}}$ in the aperture) could not be tested. However, for the maximum achievable input power of $\dot{Q}_{in} = 10\,\mathrm{kW}$, that corresponds to a part load condition of 67 %, a receiver efficiency of $\eta_{th} = 75\,\%$ at the target outlet temperature of $T_o = 900\,°\mathrm{C}$ has been obtained.

In order to characterize the thermal performance of the CentRec concept a three-dimensional, steady-state numerical model of the laboratory prototype was developed. It is based on FEM analysis and predicts receiver loss mechanisms, such as optical, radiation, convection and conduction, in detail. As the interaction of single particles is not of any interest, the modelling of discrete particles is neglected. Instead, the entire particle film is represented by continuous fluid lines, where heat is transferred due to mass transport of the fluid. The receiver rotation is considered by arranging the fluid lines on helical paths simulating the particle trajectory from a stationary point of view. Boundary conditions, such as mass flow rate, particle inlet temperature and input power were applied according to experimental parameters. The incoming heat flux distribution from the solar simulator and optical losses were calculated using ray-tracing methods.

For the evaluation of convection losses in the receiver, experimental studies were conducted investigating the effect of receiver rotation on convection. A test rig was therefore designed and tested for different rotation rates, mean wall temperatures and receiver inclinations. The experiments revealed minor influences of rotation on convective losses for the considered rotation rates. Especially for downward-facing receivers, where convection losses represent just a small part of the overall losses, the effect of rotation is less than $\pm 1\,\%$ and can thus be neglected. As comparison between experimental and literature data yielded best agreement with Clausing's correlation [18], his method was used to calculate an appropriate heat transfer coefficient, which was applied on the corresponding convective area in the numerical CentRec model.

The validation of the model with experimental data from the high flux tests reveal good agreement in the particle outlet temperature, but significant deviations in the wall temperature distribution. In a thorough sensitivity analysis a strong dependence of the modelled temperature profile on the incoming heat flux distribution was identified. Differences in the temperature profiles could be due to inaccurate modelling of the heat flux distribution as the exact simulation of the experimental conditions could not be entirely ensured. Moreover, due to the arrangement of the thermocouples along the cavity, the temperature of its back wall was actually measured while the particle surface temperature was evaluated in the simulation. Deviations between particle surface and wall temperature seem to be not negligible as the particle film thickness might play an important role. Its actual influence however could not be finally identified.

Considering the numerical model as validated as it accurately calculates the particle outlet temperature for given input power and mass flow rate, simulations of the entire load range predict a receiver efficiency of $\eta_{th} > 85\,\%$ under full load condition at $\dot{q}_{in} = 1\,\mathrm{MW\,m^{-2}}$. Decreased $\dot{q}_{in}$ results in decreasing efficiencies down to $\eta_{th} = 54\,\%$ for $\dot{q}_{in} = 0.3\,\mathrm{MW\,m^{-2}}$ at $\alpha = 90°$.

The present work could successfully demonstrate the feasibility of the proposed CentRec concept and its promising thermal performance. For future research however, the sensitivity of particle film behavior to external influences such as unbalances, eccentricity and vibration, should be studied in more detail as the film is strongly susceptible to these effects. Another interesting aspect that should be addressed is the number of actual moving film layers. The influence on the thermal performance of the receiver could not be completely investigated so far and should also be considered in subsequent research espescially with regard to larger scale receivers. Finally, for model validation purposes methods and instruments should be developed and installed in order to be able to accurately determine the actual particle film temperature.

References

[1] *ISOTECH PEGASUS Kalibrier-System (Modell 853), Manual.* 57

[2] *Microtherm Block Data Sheet.* 48, 141

[3] *Netzsch DSC 404 F1, Manual.* 61

[4] ANSYS *Mechanical User's Guide.* 72, 74

[5] *VdTÜV-Werkstoffblatt, Werkstoff-Nr. 2.4663, WB 485, 12.2009.* 46, 139

[6] *Gear Nomenclature, Definition of Terms with Symbols.* American Gear Manufacturers Association, 2005. 73

[7] *VDI-Wärmeatlas.* Springer-Verlag, 2006. 69

[8] FLUENT *Theory Guide*, November 2010. 28

[9] T. BAUMANN, C. BOURA, J. ECKSTEIN, J. DABROWSKI, J. GÖTTSCHE, B. HOFFSCHMIDT, S. SCHMITZ, AND S. ZUNFT. Properties of bulk materials for high-temperature air-sand heat exchangers. In *Proceedings ISES, Kassel (Germany)*, 2011. 139

[10] E. Bilgen and H. Oztop. Natural convection heat transfer in partially open inclined square cavities. *International Journal of Heat and Mass Transfer*, **48**:1470–1479, 2005. 17

[11] W. Blanke and U. Grigull. *Thermophysikalische Stoffgrößen*. Springer, 1989. 60, 167

[12] B.M. Boubnov and G.S. Golitsyn. *Convection in rotating fluids*, **29** of *Fluid Mechanics and its Applications*. Kluwer Academic Publishers, 1995. 18

[13] R. Buck. Spray *Manual*. DLR, 2010. 12, 79, 171

[14] W. Chakroun. Effect of boundary wall conditions on heat transfer for fully opened tilted cavity. *Journal of Heat Transfer*, **126**:915–923, 2004. 17

[15] W. Chakroun, M.M. Elsayed, and S.F. Al-Fahed. Experimental measurements of heat transfer coefficient in a partially/fully open tilted cavity. *Journal of Solar Energy Engineering*, **119**:298–303, 1997. 17

[16] Y.L. Chan and C.L. Tien. Laminar natural convection in shallow open cavities. *Journal of Heat Transfer*, **108**:305–309, 1986. 17

[17] A. M. Clausing. Convective losses from cavity solar receivers - comparisons between analytical predictions and experimental results. *J. Sol. Energy Eng.*, **105**:29 – 33, 1983. 16, 20

[18] A. M. Clausing, J.M. Waldvogel, and L.D. Lister. Natural convection from isothermal cubical cavities with a variety of side-facing apertures. *J. Heat Transfer.*, **109**:407 – 412, 1987. 16, 32, 33, 40, 41, 70, 83, 142, 150

[19] M. F. Cohen and D. P. Greenberg. The hemi-cube: a radiosity solution for complex environments. In *SIGGRAPH '85 Proceedings of the 12th annual conference on Computer graphics and interactive techniques*, 1985. 72

[20] G. Comini, S. Del Guidice, and C. Nonino. *Finite Element Analysis In Heat Transfer: Basic Formulation & Linear Problems*. Taylor & Francis, 1994. 66

[21] G. Dibowski, A. Neumann, P. Rietbrock, C. Willsch, J-P. Säck, and K-H. Funken. Der neue Hochleistungsstrahler des DLR - Grundlagen, Technik, Anwendung. In *Sonnenkolloquium Köln*, 2007. 43, 45

[22] P.K. Falcone, J.E. Noring, and J.M. Hruby. Assessment of a solid particle receiver for a high temperature solar central receiver system. Technical report, Sandia National Laboratories Paper No. SAND85-8208, 1985. 3

[23] K.-H. Funken, B. Pohlmann, E. Lüpfert, and R. Dominik. Application of concentrated solar radiation to high temperature detoxification and recycling processes of hazardous wastes. *Solar Energy*, **65**:25 – 31, 1999. 8

[24] K.-H. Funken, M. Roeb, P. Schwarzbözl, and H. Warnecke. Aluminium remelting using directly solar-heated rotary kilns. *Journal of Solar Energy Engineering*, **123 (2)**:117 – 124, 2001. 8

[25] B. Gobereit. *Theoretische und experimentelle Untersuchungen zur Weiterentwicklung von solaren Partikelreceivern*. PhD thesis, University of Stuttgart, 2014. 5, 171

[26] J.W. Griffin and K.A. Stahl. Optical properties of solid particle receiver materials i,ii. *Solar Energy Materials*, **14**:395–425, 1986. 4, 9

[27] P. Haueter, S. Moeller, R. Palumbo, and A. Steinfeld. The production of zinc by thermal dissociation of zinc oxide - solar chemical reactor design. *Solar Energy*, **67**:161 – 167, 1999. 7

[28] C.F. Hess and R.H. Henze. Experimental investigation of natural convection losses from open cavities. *Journal of Heat Transfer*, **106**:333–338, 1984. 17

[29] C. K. Ho and B. D. Iverson. Review of high-temperature central receiver designs for concentrating solar power. *Renewable and Sustainable Energy Reviews*, **29**:835–846, 2014. 2

[30] C.K. Ho, S.S. Khalsa, and N.P. Siegel. Modeling on-sun tests of a prototype solid particle receiver for concentrating solar power progresses and properties. In *Proceedings of ES2009, San Francisco*, 2009. 5

[31] G. M. Homsy and J.L. Hudson. Centrifugal driven thermal convection in a rotating cylinder. *J. Fluid Mechanics*, **35**:33 – 52, 1969. 18

[32] G. M. Homsy and J.L. Hudson. Heat transfer in a rotating cylinder of fluid heated from above. *Int. J. Heat Mass Transfer*, **14**:1149 – 1159, 1971. 18

[33] J. M. Hruby. A technical feasibility study of a solid particle solar central receiver for high temperature applications. Technical report, Sandia National Laboratories Paper No. SAND 86-8211, 1986. 3, 4, 171

[34] J. M. Hruby, R. R. Steeper, G. H. Evans, and C. T. Crowe. An experimental and numerical study of flow and convective heat transfer in a freely falling curtain of particles. Technical report, Sandia National Laboratories Paper No. SAND 86-8714, 1988. 4

[35] K. Julien, S. Legg, J. McWilliams, and J. Werne. Rapidly rotating turbulent Rayleigh-Bernard convection. *J. Fluid Mech.*, **322**:243 – 273, 1997. 22

[36] F. Kelbert and C. Royere. Study of a rotary kiln as a direct receiver of radiant energy. In *4th International Symposium on Solar Thermal Technology*, 1990. 6

[37] Y. T. Ker, Y. H. Li, and T. F. Lin. Experimental study of unsteady thermal characteristics and rotation induced stabilization of air convection in a bottom heated rotating vertical cylinder. *Int. J. Heat Mass Transfer*, **41**[11]:1445 – 1458, 1998. 19

[38] A. A. Koenig and M. Marvin. Convection heat loss sensitivity in open cavity solar receivers: Final report. Technical report, 1981. 17, 40, 41

[39] R. P. J. Kunnen, H. J. H. Clercx, and B. J. Geurts. Heat flux intensification by vortical flow localization in rotating convection. *Phys. Rev. E.*, **74**[056306], 2006. 19, 33

[40] P. Le Quere, F. Penot, and M. Mirenayat. Experimental study of heat loss through natural convection from an isothermal cubic open cavity. Technical report, Sandia National Laboratories Report, SAND81-8014, 1981. 16

[41] U. Leibfried and J. Ortjohann. Convective heat loss from upward and downward-facing cavity solar receivers: Measurements and calculations. *J. Sol. Energy Eng.*, **117**[2]:75 – 84, 1995. 17, 40, 41

[42] F. Lieneweg. *Handbuch der technischen Temperaturmessung*. Vieweg Verlag, 1976. 51

[43] D. Lin and W-M Yan. Experimental study of unsteady thermal convection in heated rotating inclined cylinders. *Int. J. Heat Mass Transfer*, **43**:3359 – 3370, 1999. 19

[44] K. Lovegrove, T. Taumoefolau, S. Paitoonsurikarn, P. Siangsukone, G. Burgess, A. Luzzi, G. Johnston, O. Becker, W. Joe, and G. Major. Paraboloidal dish solar concentrators for multimegawat power generation. In *International Solar Energy Society, Solar World Congress, Sweden*, 2003. 40, 41

[45] R. Mahoney. Thermal response of a small scale cask-like disk after heat treatment. Technical report, DOT/FRA/ORD-90/01, 1990. 23, 31

[46] J. Martin and J. Vitko. Ascuas: A solar central receiver utilizing a solid thermal carrier. Technical report, Sandia National Laboratories Paper No. SAND 82-8203, 1982. 3

[47] A. Meier, E. Bonaldi, G. M. Cella, W. Lipinski, and D. Wuillemin. Solar chemical reactor technology for industrial production of lime. *Solar Energy*, **80**:1355 – 1362, 2006. 8

[48] A. Meier, E. Bonaldi, G.M. Cella, W. Lipinski, D. Wuillemin, and R. Palumbo. Design and experimental investigation of a horizontal rotary reactor for the solar thermal production of lime. *Energy*, **29**:811 – 821, 2004. 7

[49] W. Minkina and S. Dudzik. *Infrared Thermography - Errors and Uncertainties.* Wiley, 2009. 50

[50] R. Müller. *Reaktion-Entwicklung für die solar thermische Produktion von Zink.* PhD thesis, ETH Zürich, 2005. 7

[51] A. NEUMANN. Procedures for flux measurements for solar receivers using video cameras and lambertian targets. In *SolarPACES Conference*, 1997. 55

[52] S. PAITOONSURIKARN. *Study of a Dissociation Reactor for an Ammonia-Based Solar Thermal System*. PhD thesis, Australian National University, Canberra, 2009. 17

[53] S. PAITOONSURIKARN AND K. LOVEGROVE. Numerical investigation of natural convection loss in cavity-type solar receivers. In *Proceedings of 40th Conference of the Australia and New Zealand Solar Energy Society (ANZSES), Newcastle, Australia*, 2002. 17

[54] S. PAITOONSURIKARN AND K. LOVEGROVE. On the study of convection loss from open cavity receivers in solar paraboloidal dish applications. In *Proceedings of 41st Conference of the Australia and New Zealand Solar Energy Society (ANZSES), Melbourne, Australia.*, 2003. 17

[55] S. PAITOONSURIKARN, K. LOVEGROVE, G. HUGHES, AND J. PYE. Numerical investigation of natural convection loss from cavity receivers in solar dish applications. *Journal of Solar Energy Engineering*, **133**:021004–1–021004–10, 2011. 17, 40, 41

[56] O. POLAT AND E. BILGEN. Laminar natural convection in inclined open shallow cavities. *International Journal of Thermal Sciences*, **41**:360–368, 2002. 17

[57] O. POLAT AND E. BILGEN. Natural convection and conduction heat transfer in open shallow cavities with bounding walls. *Heat Mass Transfer*, **41**:931–939, 2005. 17

[58] M. PRAKASH, S.B. KEDARE, AND NAJAK J.K. Determination of stagnation and convective zones in a solar cavity re-

ceiver. *International Journal of Thermal Sciences*, **49**:680–691, 2010. 17

[59] M. Röger, L. Amsbeck, B. Gobereit, and R. Buck. Face-down solid particle receiver using recirculation. In *SolarPACES*, 2010. 6

[60] Cha S. S. and Choi K. J. An interferometric investigation of open-cavity natural convection heat transfer. *Experimental Heat Transfer*, **2**:27–40, 1989. 17

[61] G. Saha, S. Saha, and A.H. Mamun. A finite element method for steady-state natural convection in a square tilt open cavity. *ARPN Journal of Engineering and Applied Sciences*, **2**:41–49, 2007. 17

[62] J. Schriever-Schubring. *Messung der Partikeltemperatur in einem Zentrifugalreceiver für Solarturmkraftwerke.* Master's thesis, University of Stuttgart, 2013. 51, 52, 172

[63] P. Schwarzbözl. *HFLCAL User's Guide.* 11

[64] D.L. Siebers and J.S. Kraabel. Estimating convective energy losses from solar central receivers. Technical report, Sandia National Laboratories Report, SAND 84-8717, 1984. 16

[65] N. Siegel. Private communication, 2012. 60, 71, 131, 165, 167

[66] N. Siegel and G. Kolb. Design and on-sun testing of a solid particle receiver prototype. In *Proceedings of ES2008*, 2008. 5

[67] R. Siegel and J.R. Howell. *Thermal Radiation Heat Transfer.* Taylor & Francis, 4th edition, 2002. 67

[68] CS. SINGER, R. BUCK, R. PITZ-PAAL, AND H. MÜLLER-STEINHAGEN. Assessment of solar power tower driven ultra-supercritical steam cycles applying tubular central receivers with varied heat transfer media. *Journal of Solar Energy Engineering*, **132**:1–12, 2010. 2

[69] W. B. STINE AND C. G. MCDONALD. Cavity receiver convective heat loss. In *Proceedings of the International Solar Energy Society Solar World Congress*, 1989. 17, 40, 41

[70] T. TAN AND Y. CHENG. Review of study on solid particle solar receivers. *Renewable and Sustainable Energy Reviews*, **14**:265 – 276, 2010. 5

[71] T. TAUMOEFOLAU AND K. LOVEGROVE. An experimental study of natural convection heat loss from a solar concentrator cavity receiver at varying orientation. In *Proceeding of Solar 2002, ANZSES Annual Conference*, 2002. 17, 70, 84

[72] J. R. TAYLOR. *An Introduction to Error Analysis*. University Science Books, 1997. 30, 31, 56

[73] Y. S. TOULOUKIAN AND E. H. BUYCO. *Thermophysical Properties of Matter, Volume 5*. Springer, 1970. 60, 167

[74] F. TROMBE. *Furnace for the treatment of substances by means of the energy supplied by a concentrated radiation*. US2793018, 1957. 6, 7, 171

[75] R. UHLIG. FEMRAY *Manual*. DLR, 2010. 78

[76] N. WAIBEL. *Experimentelle Untersuchung der Konvektionsverluste eines rotierenden Hohlraumreceivers für Solarturmkraftwerke*. Master's thesis, TU München, 2011. 24, 25, 32, 49, 70, 84, 171

[77] CH. WILLSCH. Private communication. 60

[78] S. F. WU AND T.V. NARAYAMA. Commercial direct absorption receiver design studies. Technical report, SAND88-7038, 1988. 2

[79] S.Y. WU, L. GUAN, J.Y. XIAO, Z.G. SHEN, AND L.H. XU. Experimental investigation on heat loss of a fully open cylindrical cavity with different boundary conditions. *Experimental Thermal and Fluid Science*, **45**:92–101, 2013. 18

[80] S.Y. WU, XIAO L., CAO Y., AND LI Y.R. Convection heat loss from cavity receiver in parabolic dish solar thermal power system: A review. *Solar Energy*, **84**:1342–1355, 2010. 17

[81] S.Y. WU, XIAO L., AND LI Y.R. Effect of aperture position and size on natural convection heat loss of a solar heat-pipe receiver. *Applied Thermal Engineering*, **31**:2787–2796, 2011. 18

[82] W. WU, B. GOBEREIT, CS. SINGER, L. AMSBECK, AND R. PITZ-PAAL. Direct apsorption receivers for high temperatures. In *SolarPACES*, 2011. 11

[83] H. Q. YANG, K. T. YANG, AND J. R. LLOYD. Rotational effects on natural convection in a horizontal cylinder. *AIChE Journal*, **34**[10]:1627–1633, 1988. 18

[84] K.C. YEH, G. HUGHES, AND K. LOVEGROVE. Modelling the convective flow in solar thermal receivers. In *Proceedings of the 43rd Conference of the Australia and New Zealand Solar Energy Society (ANZSES), Dunedin, New Zealand*, 2005. 17

[85] J.-Q. ZHONG, R. J. A M. STEVENS, H. J. H. CLERCX, R. VERZICCO, D. LOHSE, AND G. AHLERS. Prandtl-,

Rayleigh-, and Rossby-number dependence of heat transport in turbulent rotating Rayleigh-Benard convection. *Phys. Rev. E.*, **102**[044502], 2009. 19, 33

Appdx A

Measured material properties of the used particles, which are implemented in the numerical receiver model.

Absorptivity and emissivity

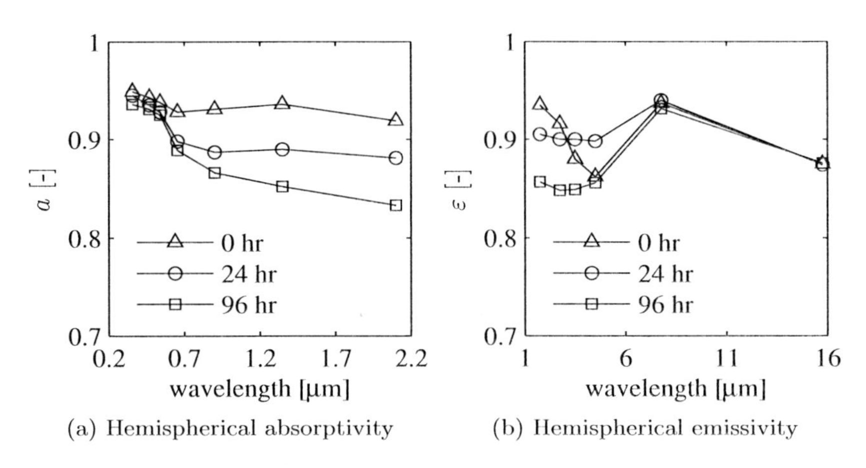

Figure 1: Wavelength dependent a and ε of used particles (Carbo HSP) after different operating hours, measured by Siegel [65].

Conductivity

To accurately determine the heat conductivity of the CentRec prototype, experiments similar to those described in Chapter 2.3.4 are conducted. A coiled heating wire is mounted in the cavity middle heating up the receiver wall to the desired temperature level. The aperture is closed by a plug made of insulation preventing convection and radiation losses into the ambient. The measured electrical power of the heating wire is then equal the conduction heat rate through the receiver at a certain wall temperature.

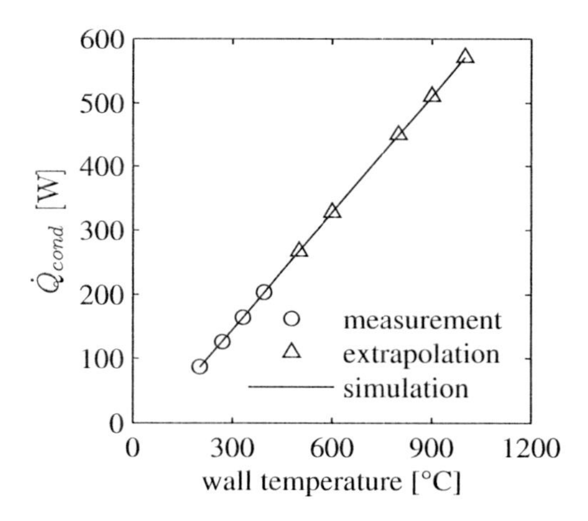

Figure 2: Comparison of measured and simulated conduction losses of the CentRec prototype. A temperature independent conductivity of $\lambda = 0.168\,\mathrm{W\,m^{-1}\,K^{-1}}$ is implemented in the numerical receiver model.

The experimental conditions are modelled in the simulation by neglecting radiation and convection and setting a constant receiver wall temperature. The resulting numerical results are then compared to the experimental measurements. An appropriate constant conductivity coefficient of $\lambda = 0.168\,\mathrm{W\,m^{-1}\,K^{-1}}$ is determined iteratively by fitting the numerical solutions to the measured data. As indicated in Figure 2 the implemented λ yields excellent agreement between measured and simulated data. Note, that due to the limited power of the heating wire a maximum wall temperature of just 400 °C are reached. However, as the measured data exhibit a linear distribution, conduction losses for higher wall temperature are extrapolated.

Heat capacity

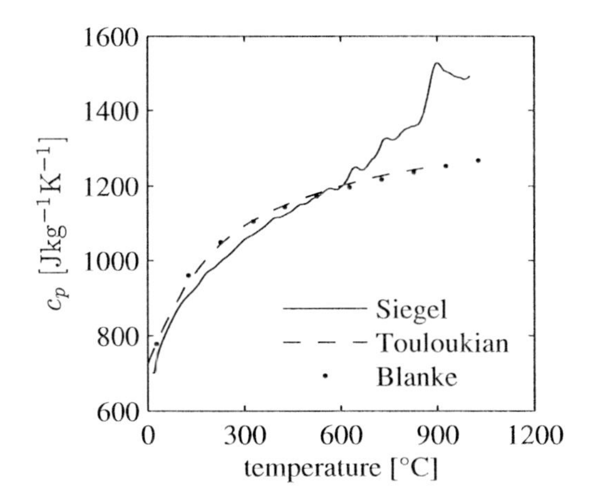

Figure 3: Temperature dependent heat capacity of used particles, measured by Siegel [65]. It is compared to the c_p of corundum, given in Touloukian [73] and Blanke [11].

Appdx B

Table 1: Uncertainties of experimental measurements.

α [°]	$\dot{q}_{in}$ [kW/m^2]	$\pm\delta\dot{q}_{in}$	$\dot{m}$ [g/s]	$\pm\delta\dot{m}$	$\overline{T}_o$ [°C]	$+\delta T_o$	$-\delta T_o$
45	265	13	9.51	0.12	385	16	15
			6.02	0.20	539	15	14
			3.94	0.12	669	35	8
			2.82	0.17	745	48	8
45	370	19	9.58	0.12	487	22	21
			6.07	0.16	688	20	14
			4.16	0.10	801	36	7
			2.93	0.17	878	52	10
90	520	26	7.30	0.21	761	25	12
	670	33	7.89	0.18	885	23	7

List of Figures

List of Tables

Publications

Parts of the thesis have been or will be published as full research articles in peer reviewed journals.

- W. Wu, L. Amsbeck, R. Buck, N. Waibel, P. Langner and R. Pitz-Paal. On the Influence of Rotation on Thermal Convection in a Rotating Cavity for Solar Receiver Applications. *Applied Thermal Engineering*, vol. 70, pp. 694-704, 2014

- W. Wu, L. Amsbeck, R. Buck, R. Uhlig and R. Pitz-Paal, Proof of Concept Test of a Centrifugal Particle Receiver, *Energy Procedia*, vol. 49, pp. 560 - 568, 2014

- W. Wu, R. Uhlig, R. Buck and R. Pitz-Paal. Numerical Simulation of a Centrifugal Particle Receiver for High-Temperature Concentrating Solar Applications. *Numerical Heat Transfer, Part A*, in press, 2015

- W. Wu, D. Trebing, L. Amsbeck, R. Buck and R. Pitz-Paal. Prototype Testing of a Centrifugal Particle Receiver for High-Temperature Concentrating Solar Applications. *Journal of Solar Energy Engineering*, in press, 2015